T0181328

Studies in Systems, Decision and Control

Volume 112

Series editor

Janusz Kacprzyk, Polish Academy of Sciences, Warsaw, Poland
e-mail: kacprzyk@ibspan.waw.pl

About this Series

The series "Studies in Systems, Decision and Control" (SSDC) covers both new developments and advances, as well as the state of the art, in the various areas of broadly perceived systems, decision making and control- quickly, up to date and with a high quality. The intent is to cover the theory, applications, and perspectives on the state of the art and future developments relevant to systems, decision making, control, complex processes and related areas, as embedded in the fields of engineering, computer science, physics, economics, social and life sciences, as well as the paradigms and methodologies behind them. The series contains monographs, textbooks, lecture notes and edited volumes in systems, decision making and control spanning the areas of Cyber-Physical Systems, Autonomous Systems, Sensor Networks, Control Systems, Energy Systems, Automotive Systems, Biological Systems, Vehicular Networking and Connected Vehicles, Aerospace Systems, Automation, Manufacturing, Smart Grids, Nonlinear Systems, Power Systems, Robotics, Social Systems, Economic Systems and other. Of particular value to both the contributors and the readership are the short publication timeframe and the world-wide distribution and exposure which enable both a wide and rapid dissemination of research output.

More information about this series at http://www.springer.com/series/13304

Kofi K. Dompere

The Theory of Info-Statics:
Conceptual Foundations
of Information
and Knowledge

 Springer

Kofi K. Dompere
Department of Economics
Howard University
Washington, DC
USA

ISSN 2198-4182 ISSN 2198-4190 (electronic)
Studies in Systems, Decision and Control
ISBN 978-3-319-87125-7 ISBN 978-3-319-61639-1 (eBook)
DOI 10.1007/978-3-319-61639-1

Printed on acid-free paper

This Springer imprint is published by Springer Nature
The registered company is Springer International Publishing AG
The registered company address is: Gewerbestrasse 11, 6330 Cham, Switzerland

To my parents and extended family.

To those working to expand the common heritage of the stock of human knowledge without prejudice.

To all the authors in the reference list and those who are not referenced but whose ideas and efforts have influenced my cognitive development and enhanced my epistemic model of statics and dynamics of the universal existence of the relational unity of matter, energy and information in one way or the other.

To all my teachers from my primary and secondary education in Ghana and my tertiary education at Temple University, Philadelphia in the United States of America.

Finally, to all those who believe that there is a general theory of unified sciences *on the basis of information and its applications through a unified practice of engineering sciences on the basis of knowledge from the general theory.*

Preface and Prologue

The linguistic term information is loaded with confusion which is enhanced by different inconsistent associated concepts such that its phenomenon has become amorphous without a head or tail. Information has been made into a strange thing, where the head is the tail and the tail is the heard. Unlike the phenomena such as matter and energy, information seems to be undefined. In this strange situation, information is taken to be anything such as data, fact, evidence, knowledge, and many others where there are some who maintain that information is information in circularity of definitional reasoning. In this confused epistemic framework, theories of data, evidence, fact, and knowledge are claimed to be the theory of information. This epistemic confusion is amplified by a number of scientists who equate communication theory with information theory. There are some who claim that information theory is a branch of applied mathematics, electrical engineering, and computer science involving information quantification. The defining claims that information is not different from the mathematical definition of a variable x, where x is x by definition in a content-free analytical environment. In this way, the variable x can be made to represent anything in any system of thought. This mathematical definition of variables constrains the development of language where classifications into categories, groups, and sets are useful in a non-content-free environment. Categories, groups, and sets impose contents on mathematical and logical variables by their definition. Within this epistemic confusion, there are claims of analytical sub-fields called the sub-fields of information theory such as coding theory, algorithmic complexity theory, algorithmic information theory, information-theoretic security, and the theory of information measurement.

The result of this epistemic confusion is the general claim that the objective of information theory is the study of the transmission, processing, utilization, and extraction of information and the resolution of uncertainties. In this respect, information theory is intimately connected, as it should be, to the theories of knowledge, decision-choice behavior, and command-control management. A problem arises when the mathematical theory of communication as advanced by Claude Shannon is taken to represent the essential component of information theory. Claude Shannon objective is to solve some specific computational problems, where information is viewed in terms of a set of messages in the possibility space, where such messages in the transmission process must pass through the probability space due to the existence of a noisy channel, and where the message sizes are to be quantitatively related to the channel capacity. In other words, what is the relationship between the message size and channel capacity in the information delivering process? This is the source-destination *information delivering problem* which is part of the general *source-destination delivering problem* of all objects and processes, where capacity must be related to size. The solution to the *size-capacity problem* requires the assumption of the given contents. It further requires that the contents be packaged in a manner that is consistent with an optimal capacity. The problem simply fits into maximization of capacity subject to size or minimization of size subject to capacity. In this way, there are size conditions and capacity conditions that must be considered in the source-destination delivery process. The conditions of the size, the capacity, or the channels cannot be used to define the contents contained in the message. This seems to be the situation in dealing with the mathematical theory of communication or information as abstracted from the tradition.

In the case of information communication, the packaging of the information contents takes the form of representation of the messaging in the form of codes and the development of optimal coding. Alternatively, the code representation of the contents contained in the message sizes that is the packaging may be taken as given constraints on the development of optimal capacities. The coding system includes the use of existing language forms or the development of new language forms that must satisfy the source-destination familiarity. Whatever may be the case of source-destination delivery optimality, the process is about controlled communication over the epistemological space. In other words, the source-destination delivery process is about communications among ontological elements in the epistemological space where cognitive agents have to work within necessity–freedom duality, and where necessity constrains freedom to generate sufficient conditions in the management of the command-control process of decision-choice systems.

In all the approaches to the theory of information, one thing that is clear is that the concept and phenomenon of information remain undefined and illusive. Information cannot be defined within the analytical system of mathematical theory or any other theory of communication, where the conditions of the probability space are of epistemic essentiality. It also cannot be defined in the analytical system of the semantic theory of information where the conditions of data space are of epistemic essentiality in the theory. It is the case that, just as the source-destination size-capacity delivery conditions cannot be used to define the contents and the phenomenon of things to be delivered, so also the size-capacity delivery conditions of communication cannot be used to define information. In all analytically conceivable space, information cannot be defined by the concepts of data, facts, evidence, and context-free representationally structural instrument such as a mathematical calculus or any element of/in the family of abstract languages. Information is an input into the communication process to deal with control-state production. In terms of the relationship between information and communication, communication is a process of action for carrying messages of information under energy while information is the essence of that which is communicated within states and over states where within states is static in nature and over states is dynamic in nature. Information, just as matter and energy, is something, and the search for this something is the preoccupation of this monograph.

Communication constituting a process of action of a carrier cannot form an input into decision-choice and transformation actions on alternatives where such alternatives must be seen as varieties with defined identities. It is that which is communicated that constitutes the essential in knowing that affects the management of decision-choice systems and that which is communicated cannot be defined by the communication process. The contents and the phenomenon of things which are being delivered cannot be defined by the means of modal conditions of delivery. Shannon's initial approach to the development of the mathematical theory of communication was intended to solve particular technical problems in relation to size-capacity conditions in the source-destination duality over the epistemological space under the command-control activities by decision-choice agents. His mathematical construct is about communication viewed as a mode of delivery of messages in a manner that produces optimal mathematical size-capacity relation. The problem over the epistemological space may be defined for person–person intercommunication, person–machine intercommunication, and machine–machine intercommunication, where social decision-choice activities are defined for transformations of socio-natural varieties. In this respect, the focus of the mathematical theory of communication was, and is, on the conditions of the efficient transmission of information when it is known through the reproduction of messages from one defined point to another specified point through noisy channels of uncertainty, where the event of the correct message being sent is measured by probability distribution over the epistemological space. This probability is part of the delivery conditions and may not be used to define the concept of information and then fix the phenomenon of information in probabilistic boundaries.

The task of the mathematical theory of communication is not on the work of the general information that contains the contents and the signals. The task seems to concentrate on how best to transmit and interpret the signal disposition given the content disposition. The theory of communication involves the delivery conditions of content disposition through signaling and optimal coding. It is part of the general theory of information and not the entire general theory of information itself. Unfortunately, the mathematical theory of communication as advanced by Shannon has helped in understanding and creating solutions to specific size-capacity problems in messaging in engineering conditions. However, the directions that have been taken by many information theorists and scientists have made the mathematical communication approach into a theory of information, where the emphasis is on the theory of quantitative measure of signal transmission and channel capacity. This mathematical approach developed on different probabilities and probability distributions has become an important constraint as well as an epistemic limitation on the development of the *general theory of information*.

The general theory of information must lead to the recognition of differences and similarities that form the development of the concepts of cluster, group, category, and set in mathematics and non-mathematical discourse to establish varieties, differentiations, and categorial varieties for comparative analysis and the understanding of conditions of static and dynamic behaviors of elements. The idea of variety is the epicenter of the development of the general theory of information and the development of forms of representation. The theory of communication deals with a small but essential part of the general theory of information, just like the theory of languages or representation deals with an essential part of communication. The general theory of information must deal with contents and transmission of the contents for which identities and categorial identities can be known in time and overtime, in such a way as to provide direct inputs into knowledge production and indirect inputs into decision-choice activities which have no material relevance without varieties that may establish comparison of alternatives.

The ideas of in time and overtime relate to the ideas of static and dynamic behaviors of varieties as revealed by the concept and phenomenon of information. The implication, here, is that the general theory of information must be seen in terms of the theory of info-statics and the theory of info-dynamics where communication is an essential part at each point of time. The epistemic frame where information is related to the concept of variety cannot be handled in any theory of communication whether mathematical or non-mathematical. The quantity of information and its measurements have nothing to do with the essential factors of information as establishing differences and commonness that are associated with varieties and categorial varieties on the basis of which socio-natural changes and transformations of socio-natural objects acquire cognitive meanings over the epistemological space. The relevance of quantity of information and its measurement, constrained by qualitative disposition, is found in the decision-choice systems for the command-control process of change and transformation of varieties as well as the assessments of true-false propositions and messages over the epistemological space.

Information communication is an important part of person-person relational complexity while information storage and retrieval is person-specific in terms of memory. Communication, storage and retrieval must be related to information content to separate relevance from irrelevance in relation to specific decision-choice action about varieties specified as alternatives. All decision-choice activities are meaningful only in the space of varieties that must be established from information stock-flow conditions. The concept and phenomenon of information must be developed to allow for variety identifications in time and overtime as well as information stock-flow conditions of varieties and categorial varieties. There is no decision and there is no choice without varieties. Even in a single element case a variety is created in decision-choice action by *dualizing* the element in terms of with and without to create two varieties for decision-choice action. The central concerns of the general theory of information are the development of solutions to the variety-identification problem, the transmission-process problem and the stock-flow problem in time and overtime.

The theory of info-statics is about the solution to the variety-identification problem in terms of definitional conditions of content, behavioral conditions of stocks and transmission relative to varieties and categorial varieties in time and in terms of the past-present existence. It is thus about an explanation of the nature of varieties and categorial varieties. The theory of info-dynamics is about the explanation of growth in information stocks of individual varieties and categorial varieties as time proceeds into the infinite future in terms of present-future existence. The epistemic force in the theory of info-statics lies in the creation of definitional conditions for defining varieties and the levels of information stocks of varieties and how this information is transmitted within the source-destination structure. The epistemic force in the theory of info-dynamics lies in the creation of logical and analytical systems to understand transformations of varieties and categorial varieties, and how new information is created as information flows to increase the information stocks of varieties and categorial varieties without the destruction of existing stocks of information.

At the level of dynamics, these are forces that bring about socio-natural transformations of socio-natural varieties and categorial varieties with the actual-potential polarities under the general principle of destruction-creation processes with the substitution-transformation dynamics, where the real cost-benefit rationality is the guide to every change. The net cost-benefit conditions register as either negative, positive, or zero-sum game for the poles in polarities through the behavior of the internal dualities. The dynamic process is such that old socio-natural varieties are faded away or destroyed losing their net benefits but retaining the information that shows their previous existence. From these destroyed varieties and loss of net benefit, new socio-natural varieties emerge with new real net benefit in replacement of the old with new information to add to the existing information. The system is such that the real old benefits are traded for new real benefits in such a way that the real net benefits of the previous variety become the real costs of the real net benefits by some socio-natural decision-choice processes with their information added to the stocks.

In this cognitive frame of forces of matter, energy, and information that create change and transformation, there is none as powerful as information. The universal system of existence is made up of nothing else but matter, energy, and information that become inputs into continual transformation destroying old forms and creating new forms in never-ending processes. Given information as establishing identities of varieties and categorial varieties, one can construct the *past-present dynamics* to understand aspects of the process of the history of categorial conversions of varieties from the beginnings. Similarly, one can construct *present-future dynamics* through decision-choice activities that can bring about continual transformation of varieties and categorial varieties of existence through the interactions of matter, energy, and information itself. The nature of universal existence is always seen in terms changing varieties through the collective interactions of matter, energy, and information and the work of energy.

In developing a general theory of information, a number of questions that must be answered come to the practical and epistemic surface. How does one know the past gone elements, present existing elements, and the future emerging elements as input-output structures of creative–destructive works of universal actual-potential polarities? In other words, how does one know the elements in the actual space and the potential space in universal existence? Do these elements exist in differences or commonness, and how do they reveal their presences, similarities, and differences? Do their natural differences and similarities change overtime and how does one know whether they change or not? In other words, how do they reveal the history of their behavior in time and overtime in terms of the past, present, and future as creatively represented by the *Sankofa-anoma* in relation to the time trinity? What are the possible relationships of the structure of these questions and possible solutions to information in time and overtime? The answers to these questions involve two organic solutions with sub-solutions that may be required for sub-problems that may arise within the organic problems.

The first solution involves the identification problem of elements leading to the establishment of conditions of identities and categorial identities with corresponding varieties and categorial varieties in time and their transforming behavior overtime. The solution to the identification problem demands the development of definitional foundations of information that will relate to identities and varieties of elements and how the information is transmitted through points of time. This is the focus of the development of the *theory of info-statics*. The epistemic epicenter of the theory of info-statics is variety that is made possible by information at a time point and helps to establish the info-stock in terms of the nature of the identities of varieties and categorial varieties at a point of time. Cognitively, the universe in relation to matter and energy has no existence without information. Language has no meaning without variety; vocabulary is devoiced of cognitive existence without differentiation and commonness; naming makes no sense without variety, and identity and rules of combination in language, as well as meaning of words and sentences, lose their communicational essence without identity that fixes variety.

The second solution involves the stock-flow problem of elements leading to the explanation and prescriptive dynamics of destruction and creation of varieties and identities, where the history of their time-point natures establishes stock-flow and its behavior overtime through the process of info-flows. The solution to this information stock-flow problem demands the construct of developmental foundations of information that will relate to the transformation of identities and varieties of elements and how their information changes and accumulates overtime. This is the theory of info-dynamics which is concerned with transformations of varieties and categorial varieties that make possible info-flows to update info-stocks in every time point. It is useful to keep in mind that a change of identity and corresponding variety is a production of new information.

This monograph is about the development of the theory of info-statics to solve the problems associated with definitional conditions of identities in relation to existences of varieties and categorial varieties. This theory follows the epistemic tradition of a number of my works on the theory of knowledge, rationality, decision-choice system, possibility–probability spaces, and uncertainty-risk phenomena, where information is taken as input into their constructs without specific definition of the concept of information that fixes the applicable range of the information phenomenon. This monograph is an initial attempt to discuss and fill the omission on the definitional aspect of information and the relevant properties of the phenomenon not in terms of communication but in terms of requirement to the understanding of the general theory of knowledge systems, rationality, and socio-natural decision-choice actions. The combined theories of info-statics and info-dynamics will show an approach to understand the claims of unified theories of physical sciences, engineering sciences, and social sciences, and how these theories constitute a unified theory of knowledge production in time and overtime. The general theory of information is also an attempt to demonstrate the existence of unified logic and mathematics that will allow a consistent development of a unified knowledge system over the epistemological space. Given the goals and objectives of the theory of info-statics, the monograph is organized in six chapters.

Chapter 1 initializes the conditions of the theory of info-statics by dealing with the relational structure between information and knowing of reality over the ontological and epistemological spaces. In this chapter, there are general epistemic reflections on the definitional problem of approaches to the concept and phenomenon of information and the lack of an explicit definition of the concept of information. The reflections lead to a number of important questions which must be answered by any theory of information. These questions relate to subjective-objective dispositions in the quantity–quality duality and how these questions relate to the universal existence in terms of matter and energy. The chapter provides a framework for points of departure from the tradition and a point of entry into the development of a general conceptual system of information theory with definition and properties.

Chapter 2 takes up the discussions on the analytical frame for the points of departure and entry and relates them to the BIT-IT problem in relation to the mind-matter problem. The conceptual foundation on the relationship of IT to BIT and vice versa is discussed leading to a search for the factors required for defining the concept of information, and then linguistically constrain its phenomenon and its uses in all areas of knowledge production as well as all areas of communication. It is in this chapter that the concept of variety, categorial varieties, and the factors of characteristic and signal dispositions are introduced and then related as a property of matter leading to the discussion of important concepts of primary category and derived category with the universal existence, and how they relate to information under conditions of acquaintance, knowing, learning, and teaching.

Chapter 3 presents the theory of info-statics, where information is argued to be a property of matter and a property of energy by categorial derivatives contrary to those who think the contrary. There is natural connectivity among matter, energy, and information, where each has a specific role to play in the static and dynamic states in the universal existence over the ontological space. This natural connectivity is made explicit with logic and exposition over the epistemological space. Matter is considered as the primary category of existence with many forms of appearance that must be recognized. This observation of different forms of matter and energy and further observations of different languages, naming, and vocabulary lead to the introduction of the definitional concepts of variety, identity, characteristic disposition, and signal disposition on the basis of which the theory of info-statics is developed with definitional foundation and extensions which must apply to all activities over the epistemological space under the conditions of cognitive actions. There is an implied theory of transmission and communication with and without noisy channels. The theory of transmission is general which is applicable over both the ontological and epistemological spaces and between them. The theory of communication is confined to the epistemological space in which the mathematical theory of communication is applicable; it is irrelevant to the ontological space. A set-theoretic framework is used to develop the definitional foundations over both the ontological and epistemological spaces. The objective of the general theory of information is stated leading to the establishment of the general information definition (GID) over the ontological space in order to obtain the concept of ontological information and its definition. The concepts of characteristic and signal dispositions are then embedded in the quantity–quality duality in a manner that allows the specification of interqualitative categorial differences and similarities as well as intra-quantitative categorial differences and similarities.

Chapter 4 is utilized to extend the ontological information to the conditions of the epistemological space, where cognitive agents operate, in order to provide a definition of epistemological information. There are discussions to show the differences and similarities between the epistemological information and ontological information. The chapter is used to introduce the conceptual comparison of the ontological and epistemological characteristic–signal dispositions and their relationship to the phenomenon of information. The concepts of epistemological variety, category, characteristic disposition, signal disposition and their relationship to theories of representation, language, and coding under methodological nominalism are discussed in relation to the definitional foundations of information. On the basis of these concepts, the epistemological information is defined as a derivative from the ontological general information definition in relation to matter and energy in terms of the concept of characteristic–signal disposition. The defined general information is then related to the conditions of acquaintance through the concept of signal disposition. It is within this epistemic structure that the concepts of exactness and inexactness are introduced and related to possibility and probability spaces to give a framework to define channel noises and their relation to defective-deceptive information structures, risks and size-capacity conditions of messages and channels over the epistemological space. These concepts are not available to us in the space of ontological actions. The chapter is concluded with a set-theoretic summary of, and reflections on, the general information definition.

Chapter 5 presents the relational structure of information, variety, category, ordering, and socio-natural transformations over both the ontological and epistemological spaces. The objective is to relate variety to alternatives in ranking and decision-choice activities over the epistemological space. The conceptual framework of conditions of informing, knowing, learning, and teaching is presented in order to answer a set of questions in the sender–receiver process over the epistemological space. The concept of variety holds in both the ontological and epistemological space. The relational concept of variety to alternative and ranking holds only over the epistemological space and not the ontological space. The set of questions requiring answers are fundamental such as who is informing? Who is informed? What is learned? What is known? How does the knower know that he or she knows? There are many other fundamental questions relating to the source-destination process. The value of information is discussed relative to the source and destination conditions in terms of symmetry and asymmetry leading to the relative measurements of symmetric and asymmetric values of information. The chapter links the concept and phenomenon of information to the existence and nature of possibility and probability spaces and sets. This linkage leads to the discussion of the relational structures of information, expectation, anticipation, surprise, and decision. The discussions are extended to the past-present-future structures of information flows, where the concepts of discount, forecasting, prediction, and prescription are defined and discussed to conclude the chapter.

Chapter 6 builds on Chap. 5 and examines some important concepts such as data, fact, evidence, evidential things, and knowledge in relation to the concept and phenomenon of information. This chapter is used to present an idea that data derivatived from the signal disposition component of characteristic–signal disposition; fact is derived from data; evidence is derived from fact, evidential things are derived from evidence while knowledge is a composite derivative from information. Some aspects of the semantic theory of information containing the standard definition of information (SDI) plus declarative, objective, and semantic (DOS) with enhancement are examined with criticism. Some questions are raised about the concept of epistemic modalities with some discussion. Algebraic definitional structures of data, fact, and evidence are presented to enhance the verbal definitions leading to the discussion on the similarities and differences among them.

The monograph of the theory of info-statics as definitional foundations begins with a preface and prologue on the nature of the set of problems of the general theory of information which is divided into the theory of info-statics and the theory of info-dynamics. It ends with reflections on theories of information, knowledge, and decision to provide an entry point into the development of the theory of info-dynamics. The conditions of the theory of info-statics initialize the dynamics of the information process.

Washington, DC, USA   Kofi K. Dompere

Acknowledgements

Thanks to all my friends, critics, and admirers and the members of the Academy who have given me intellectual encouragements and emotional support in pursuing my framework of cognition of universal existence. I am grateful to the staff of the Legon and Akuafo Halls at the University of Ghana for accommodations, when I was editing and refining my views on information. I express my thanks to my admirers and critics whose positive and negative reflections on my works have enhanced and strengthened my theoretical and philosophical convictions and made this work on information theory enjoyable to the finish. My thanks also go to all potential admirers and critics and to those who will take some time to research on static conditions of qualitative and quantitative dispositions in the definitional meaningfulness of information through the concept of characteristic–signal disposition in order to establish a distinction and similarity required for the understanding of existence of varieties and categorial varieties, where the existence of varieties and categorial varieties justifies individual preference ordering in the individual decision-choice space as well as presenting conflicts in collective preferences in the collective decision-choice space. This monograph has also benefited from the gains of methodological frontiers of the fuzzy paradigm, principle of opposites, and my discontent with the current fuzzy and exact mathematical theories on information. I thank Prof. Kwabena Osafo-Gyimah for his continual encouragement, inspiration, and motivation. Special thanks go to Ms. Jasmine Blackman for her proofreading suggestions on the first draft. Controversial ideas and terminologies are intentional, and intentionally directed to restructure the paradigm of thinking on the concept and phenomenon of information in order to establish information as a property of matter and energy as well as a methodological foundation to the understanding of the unity of knowledge-decision-choice systems over the epistemological space. I accept all responsibilities for errors in mathematics and logic that may undoubtedly arise.

Washington, DC, USA   Kofi K. Dompere

Contents

Preamble

"The purpose of what follows is to advocate a certain analysis of the simplest and most pervading aspect of experience, namely what I call 'acquaintance'. It will be maintained that acquaintance is a dual relation between a subject and an object which need not have any community of nature. The subject is 'mental', the object is not known to be mental except in introspection. The object may be in the present, in the past, or not in time at all; it may be a sensible particular, or a universe, or an abstract logical fact. All cognitive relations—attention, sensation, memory, imagination, believing, disbelieving, etc.—presuppose acquaintance" (Bertrand Russell, [R3.79, p. 127]).

"The obvious characteristics of experience seem to show that experience is a two-term relation; we call the relation *acquaintance*, and we give the name *subject* to anything which has acquaintance with objects. The subject itself appears to be not acquainted with itself; but this does not prevent our theory from explaining the meaning of the word '*I*' by the help of the meaning of the word 'this' which is the proper name of the object of attention. In this respect, especially, we found our theory superior to neutral monism, which seems unable to explain the selectiveness of experience" (B. Russell, [R3.79, pp. 173–174]).

"Scientific progress has been two-dimensional. First, the range of questions and problems to which science has been applied has been continuously extended. Second, science has continuously increased the efficiency with which inquiry can be conducted. The products of scientific inquiry then are (1) a body of information and knowledge which enables us better to control the environment in which we live, and (2) a body of procedures which enables us better to add to this body of information and knowledge.

Science both informs and instructs. The body of information generated by science and the knowledge of how to use it are two products of science" (Russell L. Ackoff, [R16.1, p. 3]).

"All economic decisions, whether private or business, as well as those involving economic policy, have the characteristic that quantitative and non-quantitative information must be combined into one act of decision. It would be desirable to understand how these two classes of information can best be combined. Obviously, there must exist a point at which it is no longer meaningful to sharpen the numerically available information when the other, wholly qualitative, part is important, though a notion of the 'accuracy' or 'reliability' has not been developed" (O. Morgenstern, [R13.18, pp. 3–4]).

"The chief thing is to understand that there is a fundamental difference (in the field of economics) between mere *data* and *observation*. The latter are naturally also data, but they are more than that. They are selected. They are supposed to arise from planned observation, guided by theory, which however need not necessarily be tied to controlled experiments. *Observations* are deliberately *designed*; other *data* are merely *obtained*. Together they constitute economic information which is related to the entire body of information, partly deriving from it, partly illuminating that section of problems that is not yet understood. Theory itself is never based solely on ordinary data in the above sense, i.e., merely obtained information with largely unknown but probably exceedingly wide error margin. Theory, moreover, is constructed and invented; data are merely gathered and collected even though this involves always administrative planning…

It is desirable to set forth systematically the relationship of such terms as 'observation,' 'data,' 'statistics,' and 'evidence.' Our use may not find general acceptance, but, in order to achieve precision, clarity of the terminology is essential" (O. Morgenstern, [R13.18, pp. 88–89]).

The quotations that have been presented in this preamble point to certain difficulties in the traditional definitions of the concept and phenomenon of information. One cannot speak of the usefulness of information without a reasonably clear definition that helps the understanding of its linkage to knowledge and decision-choice systems that find meaning in varieties over the epistemological space. Information is neither transmission nor communication, both of which are actions which are made possible by the existence of information under the conditions of matter and energy. It is this linkage system of information, knowledge, decision, choice, and practice in the universal space of varieties that justifies the reflective notion that when ignorance ascends to the throne of human organization, lies become elevated to the deputy of governance and the first casualty is truth with a violently relentless prosecution of knowledge under the principles of disinformation and misinformation. Alternatively, when knowledge ascends to the throne of human organization, truth becomes elevated to the deputy of governance and the first casualty is lies with a violently relentless prosecution of ignorance under the principles information, freedom, and justice. The confused understanding of the defining nature of the concept and phenomenon of information increases the complexity of the relationship that connects ignorance, knowledge, truth, and lies in the collection of problem-solution dualities over the epistemological space, over which cognitive agents operate to understand the necessary conditions and create the sufficient conditions in order to connect freedom to necessity, where the toolbox to deal with disinformation and misinformation is the practice of the principle of doubt at the expense of credulity.

Chapter 1
The Theory of Information and Knowing: Ontology Epistemology Reality

1.1 General Epistemic Reflections on Information and Knowledge

In this monograph, we shall concern ourselves with the morphology of non-standard information structure and its effects on the controllability of socio-natural decision-choice systems. In doing so, a number of questions tend to arise. The primary questions center on the concept and the definition of information. Given the concept and the definition of information, the derived questions involve the concepts of standard and non-standard information structures, their transformations and communications. What are the similarities and differences between standard and non-standard information structures? To identify the differences and similarities between the two conceptual structures of standard and non-standard, we need to have a clear and conceptually workable definition and explication of information and information structure. The definitional structure of standard information has been provided in a number of places [R12.18, R12.19, R12.20, R12.22, R12.23, R12.45, R20.6, R20.11, R20.12]. The definition of standard information structure takes two and interrelated paths of communication theory of information [R20.7, R20.11, R20.12] and the semantic theory of information [R12.4]. The emphasis depends on how one views the uses of information. Both definitional paths use the concept of probability. In the case of the semantic framework, Bar-Hillel & Carnap state that *the fundamental concepts of the theory of semantic information can be defined in a straight forward way on the basis of the theory of inductive probability that has been recently developed by one of us* [R12.3]. The mathematical theory of communication has an approach where the theory of communication is used as an analytical vehicle to define the amount of information as a measure of the statistical reality through the concept of probability of occurrence. In the standard-information framework, what is the definition of the concept of probability? Is the concept of probability an information-derived or is the concept of information a probability-derived? What is the relationship between

© Springer International Publishing AG 2018
K.K. Dompere, *The Theory of Info-Statics: Conceptual Foundations of Information and Knowledge*, Studies in Systems, Decision and Control 112, DOI 10.1007/978-3-319-61639-1_1

the concepts of probability and possibility and how do they interrelate within the quality-quantity duality with relational continuum and unity? Both the communication concept and the semantic concept of information of the standard information structure, seem to assume the definition and explication of the concept of information away. In other words, the phenomena of information, probability and possibility are assumed as known. In the mathematical theory of information that reflects the communication approach to establishing the definition of the concept of information, the meaning and the truth of messages are not of concern. As such, they are restricted to a limited domain of some present and future explanations of a general concept of information-knowledge production. As new problems emerge in the decision-choice system, such as *deception* and the development of the *science of deception* within the management of command and control structures, the utility of the standard approach becomes limited and sometimes analytically helpless. For example, on what basis does one claim something to be true, false, valid, real or fiction in the spectrum of socio-natural activities? The answer to this question involves comparative analysis and cannot be simply information. If one claims information as the answer, then another question follows as to what is information and how does one know that there is information? Here, it may be noted that both the mathematical theory of communication and the semantic theory of information are about communication among objects in terms of energy actions.

It is, therefore, useful to present a new framework for the definition and explication of information for the development of self-containment in the current monograph. This new framework will lead to a definition and explication of *non-standard information* and *information structure* that will make the concept of information explicit. The objective is to develop pathways to relate the information structure to controllability of socio-natural decision-choice systems through knowledge productions which must be related to the value of communication, learning and knowing. There are needs in this new framework for obtaining analytically workable general definitions of decision, choice and decision-choice system. Given a satisfactory definition of information, a number of fundamental questions tend to arise in both ontological and epistemological spaces as we examine conditions of controllability and convertibility within any system's structure. (a) At the level of ontology, how does nature produce, store, process and use information? (b) At the level of epistemology, how do cognitive agents obtain, code, store and process information to create conditions of knowing to obtain knowledge about reality as seen from exact sciences, such as physics, and inexact science such as economics? (c) At the level of natural transformations, how does nature relate and manage the relational structure of convertibility and controllability of ontological elements? (d) What is reality and how is reality related to the process of knowing? (e) Does reality present itself the same way in both ontological and epistemological spaces? (f) At the levels of physical engineering and social engineering, how do cognitive agents relate the processes of natural transformations to social transformations under conditions of general information structure?

The search for answers to these fundamental questions will begin with critical examination of information-knowledge structures and how these information-knowledge structures relate to controllability and convertibility of decision-choice systems within the ontological-epistemological polarities. From the viewpoint of knowledge production, we need an analytical clarity between the concepts of ontology and epistemology. Under the viewpoint of decision-control process of systems, we need to examine the convertibility, knowledge and credulity of the source of information. The credibility of the source of information must be examined by relating it to the phenomena of *subjective-objective duality* in the knowledge-production process. Similarly, the subjective-objective duality must be mapped onto the dualistic structure of *qualitative-quantitative dispositions*. For reasons of controllability and convertibility of states, the conditions of subjective-objective duality and qualitative-quantitative duality under the information-knowledge structure must help in instrumentation of useful control elements that will constitute the general analytical tool box for the management of commands and controls of the decision-choice system at social levels. The dynamics of the decision-choice system under commands and controls for any given object must meet observability conditions that allow the assessment of the state of the system and the distance from the preferred destination. The conditions of the preferable state are the same as the conditions of optimality which is the temporary preferred state in the epistemological space and the temporary final state in the ontological space. The distance between any state and the destination state in the ontological space will be called *ontological-controlled deficiency* which must be internally corrected by natural decision-choice actions for any given ontological information structure. The distance between any state and the preferred state in the epistemological space will be called *epistemological-controlled deficiency* which must be internally corrected by social decision-choice actions for any given epistemological information structure. What are the similarities and differences between the ontological and epistemological information structures? Similarly, what are the similarities and differences between ontological-controlled deficiency and epistemological-controlled deficiency? How can the similarities and differences in all possible cases be known?

The correction of ontological-control deficiency is an objective process which depends on an objective information. This objective information structure exists whether the ontological objects are aware or not. The correction of epistemological-control deficiency is a subjective process which depends on subjective information. The subjective information structure exists as a result of existence and awareness of ontological objects. The corrections of ontological-control deficiency and epistemological-control deficiency are information-decision-choice processes that relate to self-correcting and self-organizing control systems. The complexities of these processes are such that the definition and structure of the standard information composed of semantic and communication approaches do not constitute solutions to the tasks of problems of the general class of self-correcting systems especially when categories are under epistemic and ontological work. The transfer of definitions of standard (semantic and communication) information and information structure to the fields of non-standard information may be simply grafting on a process as well as possibly misleading the specification of the analytical structure. The problem of non-standard information structure will be formulated and discussed in reference to socio-natural decision-choice systems. For the validity of the analytical discussions and derived conclusions, it will be necessary to show that the social decision-choice systems are isomorphic to the physical decision-choice systems and that the social management of commands and controls are *ontological mimicry* at the level of biophysical decision-choice systems.

The epistemological deficiency is simply due to the nature of information and information structure as will be discussed. The concept of ontological mimicry must be defined and explicated. Additionally, the ontological mimicry process must be shown to relate to internal activities of command, controllability, observability and convertibility of states of the decision-choice system. The theory of the management of commands and controls is an information-decision-choice process and must be useful in dealing with quantity-time problems, quality-time problems and relational problems of quantity-quality duality with neutrality of time. Here, the management of commands and controls must deal with three types of equation of motions. They are the quantitative equation of motion for a constant quality, the qualitative equation of motion for a constant quantity and the simultaneity of equations of quantity-quality motion in relational unity. The unified elements of all managerial structures are the available information-knowledge structure that must be defined and explicated for each phenomenon under the decision-choice system. The concepts of command, controllability and observability are familiar. The concept of convertibility, however, may be unfamiliar, and hence requires an explanation of how it relates to other essential concepts in statics and dynamics of the decision-choice system. The importance of the concept of information-decision-choice processes that impose on any decision-choice system, the attributes of self-correction and self-organizing should not be underestimated. In fact, the behaviors of all elements in both ontological and epistemological space are governed by nothing else but decision-choice processes under information-energy actions.

The initialization of discussions on theories, conclusions and practices of information, decision and choice must be placed in the spaces of ontology and epistemology whether the phenomenon resides in a natural or social science. Controllability of decision-choice systems must be placed within the information-knowledge duality under the principle of relational continuum and unity as the guiding framework of actions for continual transformations that are true of all ontological and epistemological elements. What is the analytical relevance in introducing information-knowledge duality with relational continuum and unity in the decision-choice system? Is information not the same thing as knowledge? If the answer is no, then it will be useful to examine the defining differences and similarities that set them apart as well as unite them in the process of thought as well as their utilities in decision-choice actions. What is the relational structure of uncertainty and ignorance? Is the concept and the material meaning of ignorance derived from uncertainty, is the concept and material meaning of uncertainty derived from ignorance, or, are the concepts of ignorance and uncertainty the same? Is it meaningful to speak of *ontological uncertainties* and *epistemological uncertainties* and how are they related to the information-knowledge duality? Similar questions come to the surface in dealing with static and dynamic decision-choice systems as they are seen in transformation and change. Is the definition and explication of ignorance derived from knowledge or from information? Similarly, is the definition and explication of uncertainty information-derived or knowledge-derived? The management of commands and controls of the decision-choice system is the production of outputs (outcomes) that must have inputs. Are these inputs information or knowledge, none of them or both?

The inputs of the decision-choice system to be transformed into outputs are dependent on the processes of knowing of the universal elements to produce actions. Conceptually, it may be useful to speak of an *ontological knowing process* and an *epistemological knowing process*. In general, however, should the process of knowing be placed in knowledge-ignorance duality with relational continuum and unity where knowledge and ignorance appear in degrees of knowing whose sum, for any particular phenomenon, may be expressed as one? Similarly, should the knowledge-ignorance duality be mapped onto the information-uncertainty duality? The answers to these implicit and explicit fundamental questions that have been raised require conceptual definitions of information and knowledge, and how they may be related to ontological and epistemological realities. The ontological and epistemological realities must be connected to the process of knowing and management of systems' controllability through decision-choice actions. The control actions are defined in the quality-quantity space with neutrality of time such that there are *quality-time phenomena, quantity-time phenomena* and *quality-quantity-time phenomena*. The analytical process of these conceptual definitions leading to the meanings and awareness of the varieties and their identities defines a structure of *info-statics* and the corresponding *theory of info-statics*.

At the level of epistemology, the relational structure of these concepts in knowing is generated by an *epistemic process* which must be connected to representations such as the vocabulary and grammar under methodological nominalism which in turn must be connected to the language of science and all areas of the knowledge system through the methodological constructionism and reductionism. The theory of knowing is spun by *epistemic categories of reality* and not *ontological categories of reality*. These epistemological categories are *derived categories* under an epistemic rationality and must be related to the ontological categories which are the *primary categories*. It is the organic process of moving from the epistemological space to the ontological space to claim a *known item* (called *IT*), that information, information representation and cognitive-input processing are defined in terms of paradigms of thought which become not only necessary but imperative. The paradigms of thought function as enabling instruments that are defined for epistemic activities within the information-knowledge duality as well as the uncertainty-ignorance duality. Every selected paradigm of thought must meet the conditions of methodological nominalism, constructionism and reductionism within the *epistemological-ontological polarity* (These conditions will not be stated and explained in this monograph, but interested readers may consult [R4.7, R4.13, R17.15, R17.16]). All these questions and abstracted answers must be linked to *inputs* and *outputs* in the decision-choice processes involving the management of commands and controls to induce transformations of varieties in socio-natural systems under the principle of relational complexity. At the level of ontology, there is the organic concept of an *ontological decision-choice process* which is a set of individual ontological decision-choice processes involving onto-logical elements and ontological input-output structures within relevant environ-ments. At the level of epistemology, there is also the organic concept of an *epistemological decision-choice process* which is a set of individual epistemolog-ical decision-choice processes involving cognitive agents using appropriate inputs. The time-point conditions of the decision-choice process specify the info-static state that may be used to specify the initial conditions of *info-dynamics* and the con-struction of the *theory of info-dynamics*.

The process of knowing is the linkage between the ontological categories of reality ("the *IT*") which are created by natural processes, and epistemological cat-egories of reality ("the *BIT*") which are created by cognitive agents in the episte-mological space. The epistemic process is guided by a conceptual framework of paradigms that serves as an *information processing machine* for knowing within the source-destination duality. The understanding of the morphology of any paradigm of thought requires a clear understanding of its input of thought creation. The knowing is composed of the discovery and understanding of the nature and behaviors of ontological elements and their corresponding categories. In this respect, the act of knowing of ontological categories is the work of a *paradigm of thought* operating on *information structure* whose elements must be specified, coded and symbolically represented for an epistemic operation. The symbolic representation of information (the bit") is part of the acquaintance and language which reflect some *sense data*. The concept of data must be clearly defined, explicated and linked to the concept of information. Here, a question arises as to whether data is information or information is data. Is the concept of data information-derived, or is the concept of information data-derived? The under-standing of the information-data relational structure is important in the continual advancement of information technology and frontiers of sciences under exact and inexact nature.

The elements in the ontological space constitute the *identities* of varieties which are completely described by the *ontological information* and hence contain com-plete knowledge in their states of existence. The elements in the epistemological space constitutes *derived epistemological elements* which are described by *defective information structure,* composed of vagueness and incompleteness, which is abstracted by cognitive agents through the methodological nominalism [R3.18, R3.19, R3.34, R3.49, R3.50, R3.53, R3.54, R3.80]. The ontological categories as identities constitute a *family of primary categories,* while any category in the epistemological space constitutes a *derived category*. The theory of knowing is to establish *necessary and sufficient conditions* for an isomorphic relation between an epistemological category and an ontological category where the theory of knowing is guided by a *methodological trinity* of nominalism, constructionism and reductionism to produce a claimed known item. The methodological trinity of nominalism, constructionism and reductionism constitutes a knowing mechanism which provides justified necessary and sufficient conditions of a belief system on the basis of the known, for the management of the commands and controls in the decision-choice system of any kind. As discussed, the objective of knowing is to produce inputs into the socio-natural decision-choice systems. Over the epistemological space, the methodological constructionism and reductionism are at the mercy of methodological nominalism that must carry their inputs where such inputs must be shown to relate to information under a general definition. The social decision-choice system is composed of humanistic decision-choice systems and physical decision-choice systems, both of which are under the management of cognitive agents.

The importance of these discussions is to specify the necessary and sufficient conditions for the definition and explication of the concept of information and information structure that will be useful as inputs into the statics and dynamics of the socio-natural decision-choice systems. The relational nature of the methodological trinity of nominalism, constructionism and reductionism is presented as an epistemic geometry in Fig. 1.1. In dealing with the complexity of definitions of information and information structure at the level of epistemology, nominalism

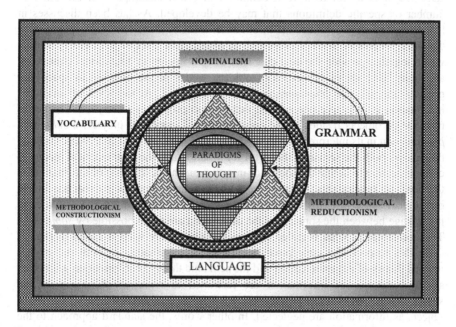

Fig. 1.1 Cognitive geometry of methodology and language in thought production

relates to language, vocabulary and grammar which allow information to be encoded and decoded, while constructionism and reductionism relate to paradigms of thought given the information and information structure to assist the decoding process. The concept of information is still not defined. The reason for a lack of a definition of the concept is simply a searching for a continual framework that will allow the understanding of the phenomenon of information and a construct of a general definition of information that is applicable in all cases over both the ontological and epistemological spaces. The search is therefore on a general framework for a general information definition that is applicable over all epistemic areas without exception.

1.2   A General Framework for Examining the Definitions of the Concepts of Information and Information Structure

So far, the concepts of Information and information structure are very broad with multiple meanings in such a way that the many definitions are restricted to needs and uses. In defining information, the general framework takes as its point of entry the idea that all these socio-natural needs and uses belong to the general class of decision-choice activities. For the general purpose of socio-natural decision-choice activities, it is useful to have a definition that will encompass all the different number of specific definitions that may be developed. As has been discusses in [R4.7], the general definition of information must be developed from interdependent components of a set of *properties of objects* and a set of *relations of objects* which together provide the dual character of information in a relational continuum and unity. The set of properties of objects resides in the ontological space and projects an*objective phenomenon* as will be argued. At the level of ontology, information is objective phenomena that present properties of objects. The set of relations of objects resides in the epistemological space and projects a *subjective phenomenon* in its interpretation as will be argued. It is this dual character of objects that allows both the semantic theory of information and the mathematical theory of communication to define the source of information and destination of information at the level of relationality. The source and destination find their existence in the ontological space in terms of material existence. The linkage and the activities in the linkage between the source and the destination are made possible by *something*. This something is what is called *energy*. The objects are what is called matter.

In this respect, the conceptual definitions of information in both the mathematical theory of communication and the semantic theory of information with other standard theories of information are restricted to relations of objects in the source-destination duality over the epistemological space where the properties of objects as information are neglected. In other words, the standard approach to the definition of the concept of information is restricted to the component of subjective phenomenon to the neglect of the component of an objective phenomenon. A general definition of information must start with the properties of objects that constitute the differential distribution of identities of varieties of matter. The properties of objects can then be transferred as subjective concepts of information and information structure. It is through the process of defining the concept of information structure, that information becomes defined in relation to specific needs in the epistemological space.

In this respect, information structure cannot constitute a vehicle to define the concept of information. Without the existence of these ontological objects, relations of objects are undefinable. The relations of information are established as subjective phenomena that give meaning to information transmission defined in terms of quantity and quality of the transmitted as seen from the source object and by the destination object. It is through the transmission process that the concepts of possibility, probability and reality arise in the epistemological space. These concepts do not arise in the ontological space [R4.7, R4.10, R4.13]. The relational structure of matter, energy and information is presented as a cognitive geometry in Fig. 1.2. It is not by accident that the dominant scope and emphasis were given to energy and not to information in the previous era of scientific advance due to the complex nature of

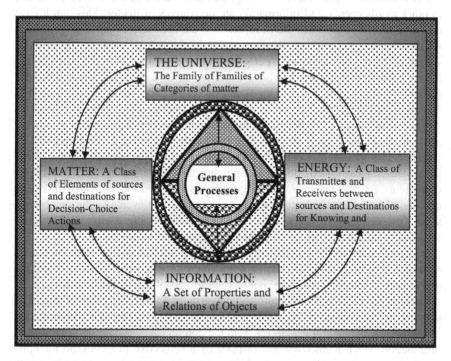

Fig. 1.2 Epistemic geometry of a relational structure of matter, energy and information for a definition of information

information and its relationship to energy in the ontological space. All that cognitive agents can do are progressive actions that are undertaken in the epistemological space to understand the information-energy relation given matter. Without the understanding of energy, it would be an increasing difficult to understand information, varieties of information structures and the identities of the ontological varieties. Here, the universe is conceived to be composed of matter and energy that are welded together by information in the sense that both matter and energy are composed of a distribution of varieties with identities that are established by information without which these identities are unknowable. The existence of matter and energy implies the existence of information the definition of which must be clearly established in generality.

The search for the general concept of information must begin with an entry point that constitutes an axiomatic foundation in defining the concept of information and an information phenomenon. This axiomatic foundation is the existence of matter and energy in relational continuum and unity in the ontological space. The matter and energy are ontologically linked by *information* that gives matter and energy the never-ending static and dynamic activities. In this respect, there is matter, energy and information which are being conceptualized as ontological trinity, where matter and energy constitute an organic ontological polarity. One finds matter in the primary existence while one finds energy and information in the derived existence. In other words energy and information are derivatives of matter. This ontological relational structure becomes the occupation of the activities of ontological objects whose objective is continual shaping and reshaping of the elements in terms of variety in the ontological space. It also becomes the occupation of cognitive agents in the epistemological space whose objective is to understand the ontological existence composed of objects, states and processes. The ontological activities in the ontological space are the works of ontological information-decision-interactive processes on behalf of ontological objects for internal management of command and controls of ontological decision-choice systems for continual transformations of varieties. The epistemic activities in the epistemological space are simply the works of epistemological information-decision-interactive processes on behalf of cognitive agents for the management of commands and controls of epistemological decision-choice systems to understand the activities and events in the ontological space. The defining attributes of ontological information are illusive and not controversial unlike those that present the epistemological information.

The processing of the ontological information in the ontological decision-choice system is a natural process in terms of the connectors between the source and the destination. These connectors are *natural paradigms* of transformation that present natural laws of relational continuum and universal unity with neutrality of time. The activities of the connectors are made possible by energy that resides within matter. The results of the natural paradigms from the ontological space present ontological information which is encoded and mapped by a subjective process onto the epistemological space as epistemological information that must be processed through decoding. The coding and the decoding are done by ontological objects in the epistemological space. Here, the ontological information is objective and not only that but it also constitutes knowledge in such a way where there is an equality between ontological information and ontological knowledge. The processing of the epistemological information also requires a paradigm of thought in transforming the epistemological information into an output which is called knowledge. The epistemological information like the ontological information is not knowledge. Necessary and sufficient conditions would have to be established for equality between epistemological information and knowledge. As is being discussed here, there is the *ontological paradigm* of natural creation that provides rules for acting on the ontological information to provide an input into the management of the commands and controls of the ontological decision-choice system. Similarly, there are epistemological *paradigms* of cognitive creation that provide rules of thought for acting on the epistemological information to provide an input into the management of the commands and controls of the epistemological decision-choice system.

The ontological paradigms are natural creations and follow the rules of nature. The epistemological paradigms are epistemic constructs that must be related to the type of information that is held as an input in the epistemological space. The paradigms of thought in the epistemological space may, thus, vary in accord with the concept and nature of information and information structure. Much of the controversies and paradoxes regarding the concepts of information and information structure are due to subjective information that presents itself in the epistemological space. Are inputs into the management of commands and controls of decision choice systems information or knowledge? Is subjective information the same as knowledge? Are both epistemological information and knowledge exact or inexact? Are uncertainties and risks the attributes of both ontological and epistemological spaces or are they only attributes in the epistemological space? Are the concepts of possibility, probability and actual phenomena of epistemological information combined with human ignorance? What does risk mean and connote as it is linked to the ontological space and epistemological space? It will become clear that if ignorance is lack of knowledge and risk is the presence of ignorance then risk is a phenomenon defined in the epistemological space and is undefinable in the ontological space.

The answers to these questions require a comparative analysis of the attributes of information in ontological space and epistemological space no matter of the definition provided. The definitions of the concepts of information and information structure that combine objective and subjective phenomena have not been given yet. However, it has been stated that information is composed of properties of ontological objects and relations of ontological objects, where the ontological objects constitute the sources and destinations of information transmission through activities in the encoding-decoding duality with relational continuum and unity where energy is an enabler. The comparative analytical structure of ontological and

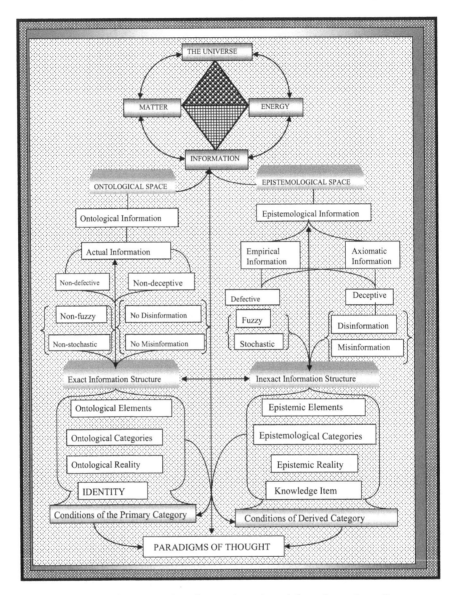

Fig. 1.3 The relational structure of ontology, epistemology, information and paradigm

epistemological information types is provided in Fig. 1.3 which presents the relational structure of ontological information and epistemological information. One important way to view this comparative structure is that the information in the ontological space represents the conditions of the *IT* while the information in the epistemological space represents the conditions of the *BIT*.

1.3   Some Reflections on the Standard Information Definition (SID)

It is now useful to turn an attention to the definition of information and the structure of information after which attentions will be devoted to the paradigms of thought in the epistemological space. Epistemic reflections on the main ideas of the conceptual definition of information in the *standard information theory* which involves the definition and structure of information will be useful in providing a point of entry for definition of the concept of information in the non-standard theory of information which is composed of interdependent sub-theories of the theory of info-statics and the theory of info-dynamics. The concepts of information and knowledge were introduced in the previous sections without their explicit definitions. It seems that in most theoretical and empirical works in exact and inexact sciences over the epistemological space, information and knowledge acquire interchangeability without specifying the conditions of their substitutability and transformation. In this way, knowledge is information and information is knowledge generating the concept of information-knowledge equality. The information-knowledge equality has become important conditions of the standard definition of information. As has been pointed out in the previous sections, the information-knowledge equality holds only in the ontological space in which case there is no uncertainty, ignorance, surprise or risk as is represented in Fig. 1.3. The knowledge production where information-knowledge equality does not hold takes place in the epistemological space. The similarity and differences will be discussed under the sub-theory of info-statics of the *theory of non-standard information*.

It is useful to state what the concept of information is not, by keeping in mind the process-path to knowledge discovery. The first thing is the identification of a phenomenon that may lead to naming and creation of a concept in terms of vocabulary within a particular language through the principle of acquaintance to generate *experiential information structure*. The next step is to clean the concept from what it is not and set the framework for its definition and possible explication. Given the definition and explication, it becomes useful to examine the content of the concept (semantic containment, quality). This process may be placed in methodological nominalism. The content of the concept may be subjected to conditions of measurement and unit of measurement (quantity). In general, one cannot meaningfully discuss the content of an unknown concept as well as measure a content of an unknown concept. This is the *quality-quantity problem* of all concepts when they are given definitions and explications. The use of the concept in any language may project deception or non-deception whose analysis as the definitional structure of non-standard information will be discussed.

The steps for the standard information definition may be presented in a cognitive geometry as in Fig. 1.4. It is useful to observe that the phenomenon of information is missing in the establishment of the process of the standard information and its theoretical development. In this respect, the standard theory of information deals

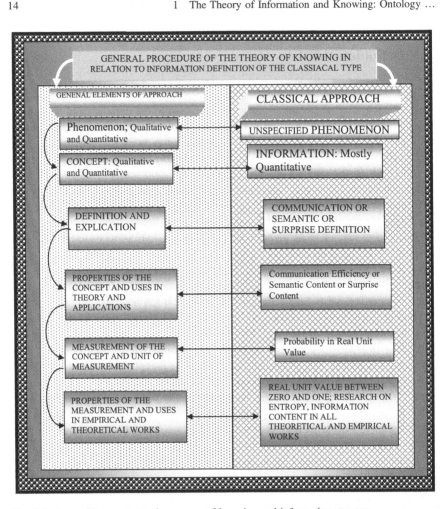

Fig. 1.4 A cognitive geometry the process of knowing and information structure

substantially with the quantity side of information. Even the semantic approach to information is also guilty of this emphasis on quantitative disposition. The question that may be asked is: in what process can the concept of information incorporate elements of quality and quantity of the phenomenon of information in the standard approach and utilize the epistemic machine of the classical paradigm of thought? The definition of the concept of information in information theory no matter how the theory is constructed cannot claim independence from the theory of knowing which may affect the form and the structure of the concept of information as conceived in the epistemological space. Analytically, it will become clear that both the classical paradigm of thought and the corresponding theory of knowing have substantial influence on the standard definition of information.

1.4   The Theory of Knowing and Information Definition in the Classical Approach

Since the definition of the concept of information cannot be devoiced from paradigms of thought and the corresponding theories of knowing over the epistemological space, it is useful to examine some questions of importance in the *theory of knowing* that may help the general definitional structure of the concept of information. The theory of knowing is directed to answer the following class of fundamental questions in cognition. (1) What does the knower know (content)? (2) What is the motivation to know (curiosity and survival)? (3) What is the method and procedures for knowing (methodological constructionism composed of laws of thought, thinking and reasoning)? (4) How does the knower know he or she knows what is claimed to be known as well as convince others that something is known (methodological reductionism composed of verification, truthfulness and communication)? (5) What is the usefulness of what is known (the utility of the known)? This class of fundamental questions of the theory of knowing is different but intimately connected to the theory of learning which must answer the class of the following fundamental questions: (1) What does the learner learn (the content)? (2) What is the motivation to learn (curiosity and survival)? (3) What are the methods, techniques and processes of learning (methods and techniques)? (4) How does the learner know that he or she has learned the content of what is to be learned (content verification and assurance)? (5) How useful is that which has been learned (the utility of the learned). The theory of knowing is the foundation of all knowledge-production processes to establish contents of information. The theory of learning is to ensure the continuity of the known through the communication of the messaging systems of the contents of information. Both theories of knowing and learning are mutually connected. They both relate to information and knowledge. Information and knowledge are phenomena, while knowing and learning are processes. The theory of knowing is directly connected to the definition and content of information. On the other hand, the theory of learning is connected to communication and messaging systems and hence a process derivative of knowing. It should be noted that both knowing and learning processes take place over the epistemological space as a derivative of the ontological space.

Every phenomenon is defined in quality-quantity space. However, it is always the case that a concept of a phenomenon may be introduced into the knowing and learning processes to capture its *qualitative existence* without a known process of measuring the qualitative content. It is within the quality-quantity duality with relational continuum and unity for every phenomenon, that the concepts of qualitative and quantitative dispositions arise in such a way that behind every qualitative disposition there is a quantitative disposition and vice versa establishing a duality with a relational continuum and unity. Information as a phenomenon resides in the quality-quantity duality with relational continuum and unity. The attributes required to define it must contain the elements of qualitative disposition and quantitative disposition. Here, the semantic containment is of a qualitative nature and is defined in a penumbral region of clarity in the space of subjective disposition which relates to qualitative disposition. It is at this point that problems of objectivity and subjectivity of information arise in the information definition, especially in the epistemological space where epistemic truth is constructed and may acquire important epistemic or alethic modalities. In duality with a relational continuum and unity, there is the concept of *dualistic difference* that has an analytical utility in terms of examining the concepts and computational conditions of *dualistic dominance, equilibrium disequilibrium, order, disorder, certainty, uncertainty and many more of these in both theory of knowing learning and decision-choice processes.*

The phenomenon of information is unspecified in the standard information structure and hence a number of theoretical difficulties arise. Since the information phenomenon is unspecified, one is not sure of the *concept of information* on which the standard information theories have been developed. From the general structure of the theory of knowing, the definition of the concept of information must not include the transmission, communication, and the measurement of its content. The information structures defined by using the concepts of transmission, semantics, and communication and content measurements belong to a class that is here referred to as the *standard information structure* otherwise, it will be referred to as the *non-standard information structure*. In general, there are theories of information codding, transmission, communication, measurement and others which are simply sub-theories of the general theory of information which is intended to examine the phenomenon of information and information-production in quantity-quality duality as well as objective-subjective duality, both of which exist with relational continuum and unity in the sense that for every dual, there is opposing dual where for example, for every objective disposition, there is a supporting subjective disposition in the duality. This is not different from the theory of knowing which is intended to examine the phenomenon of knowledge and knowledge-production. What are the questions that must be answered by a general theory of information? The theories developed around the *standard definition of information* (SDI) and the corresponding information structure using the *classical paradigm of thought* belong to a class which is, here, referred to as *classical information theory*. It may also be referred to as *standard information theory*. The theories developed around the non-standard definition of information (NSDI) and the corresponding information structure using the *non-classical paradigm of thought* belong to a class which is, here, referred to as *non-classical information theory*. It may also be referred to as *non-standard information theory*. The non-standard information theory is composed of the *theory of info-statics* and the *theory of info-dynamics*. The current monograph is devoted in dealing with the theory of info-statics.

It is useful to point out that information representation cannot be separated from information definition. The relational structure of definition and representation is made complex since they affect paradigms of thought with their logic and mathematics which constitute the epistemic algorithms for information processing given its definition and representation. Every mathematical space, just like every logical space, is information defined without which there will be conditions of epistemic viciousness. It is the information structure that allows the claim of a non-empty mathematical or logical space. Every mathematical space claims its name from the nature of information structure imposed. Similarly, every algorithm is developed in accordance with the information structure. These mathematical and logical spaces are developed in the epistemological space to process epistemological information when it is defined and represented. It is with this understanding that the paths of classical and fuzzy paradigms of thought are shown in Fig. 1.5. The epistemic differences and similarities between the two paradigms are revealed by the claims of objectivity and exactness of the information structure in the epistemological space.

There are two paradigms of thought that will be of interest in terms of their roles in information-knowledge theories. They are the classical and fuzzy paradigms of thought which constitute analytical processing machines of information and the activities in the decision-choice systems under the principle of complexity in the epistemological space. This principle of complexity is an epistemic complexity

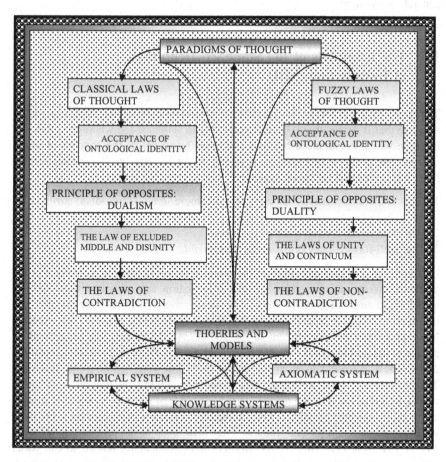

Fig. 1.5 Epistemic geometry of classical and fuzzy paradigms of thought in relation to information definition, representation and algorithm

which is a small part of universal complexity. These paradigms of thought have similarities and differences. The fuzzy paradigm can be used on the standard information structure as well as the non-standard information structure since it incorporate conditions of objective and subjective phenomena as well as qualitative and quantitative dispositions. The classical paradigm cannot deal with non-standard information structure that contains subjective and qualitative dispositions. The comparative structure of the classical and fuzzy paradigms of thought is provided as a continuity of Fig. 1.5. Here, it is assumed that the information structures become inputs into the epistemic processing machines which are generated by paradigms of thought with corresponding information-representation, laws of thought and mathematics. So far, the definition of the concept of information and its phenomenon have not been provided. It is useful to first start with the standard definition of information and analyze its strengths and weaknesses. This understanding will provide an entry point into the development of the theory of non-standard and general information.

1.5   The Essentials of the Standard Definition of Information (SDI)

The standard definition of the concept of information proceeds from a signal or message that is declarative (D), objective (O) and semantic (S) where the DOS is a message which is enhanced by *data* and *meaning* to define the concept of information. The attribute of information phenomenon as seen by the proponents are three interconnected relational structure.

1. The DOS Information contains one or more *data* (multi-data attribute of $n \geq 1$).
2. The data in the message are *well-formed*.
3. The well-formed-ness of datum is *meaningful*.

The initial standard definition of the concept of information as contained in the semantic theory of information, is that information is a message that contains well-formed and meaningful data. The declarative objectivity means that the information exists independently of the encoding-decoding process. Information, however, cannot be independent of the source and the destination which will be related to the problem of the *IT-BIT* phenomenon. The *IT* refers to existence and the *BIT* refers to representations in the processes of knowing, learning and teaching. The attributes of data, well-formed-ness and meaningfulness present some theoretical difficulties as they are referred to the existence of subjective-objective duality and quality-quantity duality. The difficulties concern the concepts of data, well-formed-ness and meaningfulness. They also relate to the problem of "*IT from BIT or BIT from IT.*" This *IT-BIT* problem is fundamentally related to categorial existence in the *primary-derived duality* and in the general reference to *actual-potential polarity,* as seen in terms of activities of production and transmission within the source-destination duality with relational continuum and unity. It is in this *IT-BIT* relational structure that the study of the problem of either epistemic or cognitive computing acquires scientific necessity.

One thing that is clear is that there will be no source and destination without the *IT*. Every *IT* is simultaneously a source and receiver not of the same message but of different messages that the *IT* may be aware or unaware of where the messages are moved through encoding-decoding duality with relational continuum and unity. Similarly, there will be no *BIT* without *IT*. The relational continuum and unity project the notion that the source is a destination and the destination is a source. In other words, the process is from *IT-to-IT* in the sense of two different *ITS* with two different messages ($IT_1 \rightleftarrows IT_2$). The *BIT* is a representation produced in the epistemological space and there will always be *IT* if there is no *BIT*. The SDI has to deal with problems of objectivity, subjectivity, quality, quantity, categorial identification, misinformation and disinformation, falsity and truth. An attempt to resolve some of these problems has led to the revision of the SDI (RSDI) [R12.3, R12.4, R12.8, R12.19, R12.20, R12.22] where the SDI is enhanced by a fourth element of truthfulness. Thus RSDI is declarative, objective and semantic if the following attributes are there.

1. The DOS Information contains one or more *data* (multi-data attribute of $n \geq 1$).
2. The data in the message are *well-formed.*
3. The well-formed-ness of datum is *meaningful.*
4. The meaningful, well-formed data are truthful.

The *semantic theory of information,* unlike the *mathematical theory of communication* and the *computational theory of information* provides a definition of the concept of information as well as an explication of the definition that may be useful for scientific analytics.

The search for a general definition of the concept of information and what it represents, however, encounters a number of difficulties within the DOS-RSDI path of conceptualizing information. Is it useful to use *data and* the *message* to define information? Is the message not the *BIT* and it not projected from the *IT to IT*? What is the concept of the message and what is the concept of data? Is the concept of data not being used to define the concept of information in the message? Given the concepts of the message and data, what are the epistemic modalities of meaningful, well-formed and truthful? Are the epistemic modalities not defined over quality-quantity duality and mapped onto objective-subjective duality with relational continuum and unity? When one examine the descriptive notion of *declarative, objective and semantic,* one may find that the concept of objectivity in DOS presents another difficulty. The existence of the concept of objectivity in the epistemological space requires an assumption of absolute existence of either truth or falsity in order to allow the application of the classical paradigm with excluded middle, where contradiction is not accepted in truth-value decisions. The same assumption does not arise in the ontological space which is the *identity*. How does the declarative, objective and semantic (DOS) information complemented by a revised standard definition of information (RSDI) allow for the analysis and claim of *reality* and deal with the *potential* where the concept of reality and potential are information-defined?

The DOS-RSDI basically deals with communication from one cognitive *IT* to another cognitive *IT*, where the transformation vehicle is the *BIT* which is powered by *energy* from the epistemological space to the ontological space. The introduction of the *cognitive IT* is through the concepts contained in the DOS-RSDI. The definition of the concept of information contained in DOS-RSDI does not allow relational structures to be analytically established between non-cognitive *IT* and cognitive *IT* or between non-cognitive *IT* and non-cognitive *IT*. It restricts the theory of knowing around cognitive agents through information-decision-interactive processes. The general definition of the concept of information must be such that the relational structure must allow knowing and learning among all *ITs* in the ontological space. An important conceptual difficulty in DOS is the concept of "objective" which is hard to maintain without qualification in the epistemological space. Both SDI and RSDI have concepts such as well-formed and truthfulness that present some difficulties. Given the DOS-SDI, the measure of the amount of information is developed on the basis of the theory of inductive probability [R12.3, R12.4]. The inductive probability cannot be separated from the claim of epistemic truth established by the classical paradigm under the principle of excluded middle where simultaneous existence of opposites in an element such as a message is denied. The implication here is the classical logical position where simultaneous existence of true and false is denied.

1.6 Mathematical Theory of Information and Communication

It is useful at this point to turn an attention to the mathematical theory of communication as another dimension of the classical information theory which also includes the semantic theory of information. The *mathematical theory of communication* which has become the *information theory,* on the other hand, takes as given some notion of the concept of information. The concept of information is not defined by but associated with the amount of information contained in outcomes or a signal under conditions of uncertainty which are measured in the probability space. The appeal to probabilistic and probabilistic measure of uncertainty allows a measure of a unit of information and a logical computability of the *expected value* of the information contained in a message or the outcome. In this frame, the value of information must be interpreted as the content of information contained in the message which must be made explicit. The fundamental problem of the mathematical theory of communication as seen by Shannon is not on the definitional concept of information, but about the amount of information contained in a transmitted message given the *concept of information* and its *content.* As pointed out by Shannon:

The fundamental problem of communication is that of reproducing at one point either exactly or approximately a message selected at another point. Frequently the messages have 'meaning'; that is they refer to or are correlated according to some system with certain physical or conceptual entities. These semantic aspect of communication are irrelevant to the engineering problem. The significant aspect is that the actual message is one 'selected from the set' of possible messages [R20.11, Vol. 27, p. 379 of (pp. 379–423, 623–656, July, October, 1948]

By intentionally neglecting the definitional problem of information, Shannon also disregarded the problem of the qualitative disposition of information within the time continuum. Information is turned into a particle and studied as a space-time phenomenon. The focus, therefore, is centered on the quantity-time problem of information, where information is quantitatively represented as an indirect or direct variable x with $P(x) = p$ in transmission and is projected into a mathematical space to be subjected to mathematical algorithms and theorems. The quantity-time problem has become the main focus of information theory, informatics and *data-matics* including data creation and data analytics, cluster analysis and pattern recognition. By sweeping away the quality-time problem of information, the information is treated as quantitative variable accorded with objectivity and placed in an exact mathematical space. The information, therefore, can be studied as a quantitative relationship between sources and destinations with enablers of trans-mitters, channels and receivers that allow analysis of velocity and accelerator. This will be referred to as source-destination polarity with activities made possible by the conflicts between encoding-decoding duality with continuum and unity whose representation contains signals. The geometric structure is generally represented as in Fig. 1.6.

It is this process of transformation by neglecting the definition of the concept of information as well as neglecting the side of qualitative disposition of information, that leads to the development of the semantic theory of information with **its**

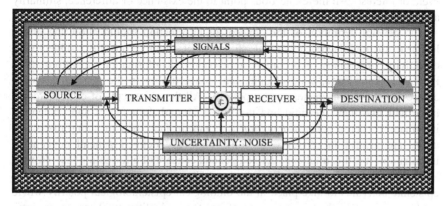

Fig. 1.6 An epistemic geometry of relational structure between transmitter-receiver duality and source-destination duality under conditions of Declarative, Objective Semantic (DOS)-Standard Definition of Information (SDI)

criticisms [R12.3, R12.4, R12.20, R12.22], and the amplification of the mathematical theory of communication into statistical theory of information and communication and the related subject areas [R8.1, R8.5, R8.8, R8.16] with a wild range of applications. It is useful at this point to bring into focus the criticism of the mathematical theory of communication by Bar-Hillel and Carnap.

> The Mathematical Theory of Communication. Often referred to also as Theory (of Transmission) of information, as practiced nowadays, is not interested in the content of the symbols whose information it measures. The measures, as defined, for instance, by Shannon, have nothing to do with what these symbols symbolize, but only with the frequency of their occurrence. The probabilities which occur in the definientia of the definitions of the various concepts in the Communication Theory are just these frequencies, absolute or relative, sometimes perhaps estimates of these frequencies.

> This deliberate restriction of the scope of Statistical Communication Theory was of great heuristic value and enabled this theory to reach important results in a short time. Unfortunately, however, it often turned out that impatient scientists in various fields applied the terminology and the theorems of Communication Theory to fields in which the term 'information' was used presystematically, in a semantic sense, that is, one involving contents or designate of symbols, or even in a pragmatic sense, that is, one involving the users of these symbols. There can be no doubt that the clarification of these concepts of information is a very important task. However, the definitions of information and amount of information given in present Communication Theory do not constitute a solution of this task. To transfer these definitions to the fields in which those semantic or pragmatic concepts are used, may at best have some heuristic stimulating value but at worst be absolutely misleading [R12.4, pp. 147–148].

Besides these criticisms, a number of theoretical difficulties arise in the mathematical theory of communication whose amount of information is valued along the concept and measure of probability. Does the concept of probability with the measure of probability relate to uncertainty? How is uncertainty defined? Does uncertainty relate to information? What does the concept of event mean; how is it related to the concept of outcome and how will it be known that the outcome is the event of interest? All these questions point to the existence of difference, similarity and commonness in conceptualizing the phenomenon information.

The general definition of the concept of information should allow the establishment of multilateral relational structure of quality, quantity, subjectivity, objectivity, uncertainty, expectation, anticipation, disinformation, misinformation, ignorance, knowledge, controllability and convertibility in the general decision-choice system within matter-energy polarity. All these concepts of critical importance will find their meaning and usefulness around the concept and content of information when a *general definition* is found. In fact, their meaningfulness and utility are always in reference to some implicitly defined general concept of information. It is this implicitly defined general concept of information that allows a linkage between *IT and IT* through the *BIT-process*. The BIT-process allows epistemic activities to take place in the ontological-epistemological polarity where conflicts are established in a system of epistemic dualities such as uncertainty-certainty duality, ignorance-knowledge duality objective-subjective duality, constructionism-reductionism duality, true-false duality, correct-incorrect duality, existence-non-existence duality, objective-subjective duality, quality-quantity duality and many others that are of occurring interest in the management of command and control of the general decision-choice system. In this respect, information constitutes a primary category on the basis of which most of these concepts are derived, and on the basis of which matter and energy are known and distinguished. The DOS-RSDI of the concept of information does not seem to allow a smooth connectivity of these concepts and uses. It is also very difficult to examine behaviors in information-knowledge duality and claim knowledge as an *epistemic reality* which must then be compared with an ontological reality. In other words, what role does one assign to the methodological duality of constructionism and reductionism and the process of establishing primary category with the derived categories and the categorial movements among them? The general definition of information must deal with the concept of information that contains the contents and communications. The distribution of the contents must establish the distribution of identities that define the distribution of varieties where the distribution must point to family of family of categories for any given point in time. The approaches of the semantic theory of information and the communication theory of information do not seem to offer a convincing solution to the task of a definition of phenomenon and related concept of knowing. A new epistemic framework is required where matter, energy and information are relationally connected in continuum and unity to establish a general information definition (GID). The epistemic framework for achieving this general information definition is what is referred to in this monograph as the *theory of info-statics* to which an attention is now turned. It will become clear that any attempt to separate information from matter and energy is simply an epistemic fiction.

Chapter 2
The Non-standard Approach to the Theory of Information: Reflections on Definition and Measurement of Information

2.1 Reflections on Points of Entry and Departure

The mathematical theory of communication and semantic theory of information have been reflected on. Any one of them, while discussing an aspect of information, cannot constitute an approach to a complete theory of information. The two together still lack the elements that will create completeness of the information theory. There seem to be conceptual and logical difficulties in both approaches that place limitations on their applicable domain especially when the process of knowledge production is contemplated between the ontological and epistemic spaces. At the level of epistemology, the analytical interest of the mathematical theory of communication centers on the problems of measurement and transmission of some undefined concept of information and the rate of transmission which are related to events, outcomes and objective probability. Similarly, at the level of epistemology, the analytical interest of the semantic theory of information centers on the problems of content and transmission of the content where the semantic attributes are used to define the concept of information and then measured with inductive probability.

The two approaches do not seem to offer a pathway that will allow an examination between the theory of information and the theory of knowledge. At the level of epistemology, how is information transmission related to the theory of knowing and then to the theory of learning? Is the measure of the content of the information related to the measure of degrees of learning and knowledge, and hence the growth of stock of knowledge? How do the mathematical theory of communication and the semantic theory of communication provide channels of linkage between the epistemological space and the ontological space? It has been pointed out in Chapter 1 that the universe may be seen as composed of matter, energy and information in relational continuum and unity which is represented in Fig. 1.2. Matter, energy and information are essential components for universal existence. The interesting thing among these three categories of existence is that one of them constitutes the primary category of existence, and the other two constitute derived categories of universal

© Springer International Publishing AG 2018
K.K. Dompere, *The Theory of Info-Statics: Conceptual Foundations of Information and Knowledge*, Studies in Systems, Decision and Control 112, DOI 10.1007/978-3-319-61639-1_2

existence that provide foundation for theories of knowing and learning. In matter resides energy and information in an important logical continuum and unity that lead to an important epistemic question regarding the epistemic primacy of existence. Here, there is *matter-energy polarity*, where matter and energy are inseparably connected by information. This is the *principle of universal connectivity* between the *matter pole* and the *energy pole* in *universal transformations* within the global *actual-potential polarity* with neutrality of time no matter how it is constructed. Without the principle of universal connectivity, transformations are impossible and the universe will be in a complete static state. The actual pole is a family of categories of actual varieties while the potential pole is a family of categories of potential varieties. In other words, the universe is conceived as composing of varieties and family of categorial varieties. It is through this understanding that the studies of categories find epistemic meaning and utility.

At the level of epistemology related to the paths of informing, knowing, teaching and learning, information must be connected to methodological nominalism which gives rise to the *universality of language development broadly defined* over the epistemological space. It is this universality of information-language connectivity that allows *encoding*, symbolism, decoding and new vocabularies to be developed also over the epistemological space. Again, this principle of *universality of information-language connectivity* cannot be established with either the mathematical theory of communication or the semantic theory of information. In this respect, learning and knowing will be difficult or impossible which then will make the creation of storage-learning machines out of reach of engineering. What then is information that allows the establishment of the relevant principle of universal connectivity? This requires a global definition of information. The global definition of information must meet the requirements of all the relevant *principles of universal connectivity* among all ontological elements.

These universalities are spanned by the role of information as a connecter between matter and energy that provide the complex analytical framework and distinctions among important concepts, such as quality, quantity, statics, and dynamics. It must be kept in mind that every phenomenon is defined by qualitative and quantitative dispositions that must be captured under the use of methodological nominalism. While the general definition of information must be neutral to events, processes, states, channels of communication, measurement, semantic structure, modes of transmission, encoding, decoding, and content validity, as well as relevance of application as seen by ontological objects, it cannot be independent of qualitative and quantitative dispositions. Both quality and quantity are attributes of matter in different states of being such that for any qualitative appearance of matter, there is a supporting quantitative disposition of matter in dualistic existence with relational continuum and unity under conditions of categorial conversion of varieties and categorial varieties [R17.15, R17.16, R17.18]. In other words, the general definition of the concept of information must be neutral from decision-choice systems but must derive its structure from the dualistic existence of qualitative and quantitative dis positions of all phenomena. It is because of the search for a general definition of the concept of information that a *non-standard approach to the theory of information* is being undertaken to provide a definition which will establish its phenomenon and content. The development of its analytical complexity is what is called *a general theory of information*. The general theory of information is made up of two parts. Part one is the *theory of info-statics* and part two is *the theory of info-dynamics*. This monograph is devoted to the development of the theory of info-statics that will make its subject content explicit. The development of the theory of info-dynamics will be undertaken in a separate volume which will also make its subject content explicit to show the differences and similarities.

2.1.1 The **BIT-IT** *Problem as a Visitation of the Mind-Matter Problem*

The mathematical theory of communication and the semantic theory of information are here classified as the classical theory of information. The analytical limitations and conceptual difficulties of this classical theory of information has been pointed out. These classical theories of information are other indirect attempts to approach the debate on the mind-matter problem where *BIT* is seen in terms of mind and *IT* is seen in terms of matter in the *IT-BIT* definitional process to the concept of information. Here, the mind is seen as some immaterial information that finds residence in the material brain which is seen as an information processor under some source-destination dualism with an excluded middle. The BIT in this respect is associated with the mind while the IT is associated with matter in a separable relation where the *BIT* dominates the *IT*. Is the mind not the result of critical and complex organization of matter? When the BIT in the mind and IT as matter are viewed in dualistic form, the emphasis is on the mind and hence on the BIT. That is to say, the information is a product of the mind and hence independent of, and fundamentally different from matter. If the mind is an immaterial information, then what is the concept of information? Is information a defined or an undefined phenomenon? Is information impinging on the mind from within the mind or impinging on the mind from outside the mind. The general definition of information must make clear whether the immaterial information is a perceptive reflection of ontological reality or the ontological reality is a derivative of immaterial information in which case it may be concluded that reality exists in the mind, which in this case, suggests that the mind constitutes the primary category of existence where, all other existence are its derivatives. Is this perceptive reflection not subjective and to what extent does it meet conditions of some degree of objectivity? It will become clear that the perceptive information has a continual limitation in all forms of capacity such as collection, retention and processing.

Corresponding to the BIT-IT or mind-matter problem is the question of what is reality and how is reality known. Does reality have qualitative and quantitative dispositions such that they mutually define their existence within the reality? It is this mutual existence that points to the establishment of the scientific area of the study of measurements, where from quality, a quantity is derived in support of qualitative dispositions, and where, from the quantitative disposition, studies of mathematical spaces and algorithms become scientifically meaningful and useful. If information resides in the mind, how does one find a measure for its content at any relevant decision-choice point? If the universe is continuously creating new information by rearranging existing matter to create new forms, can one say that these new forms are mental reflections and not ontological, and that the claim of reality is valid because of the mind as the immaterial information in the brain, and hence these new forms are defined by some mental images? Just as matter and energy exist in continual transformation in the actual-potential polarity so also is the information as a connector, where a category of information types emerges to keep the connectivity at the level of ontology. Information is indestructible but under continual growth and accumulation to update the information stock which is nothing but accumulated history. This is the *principle of information indestructibility* that supports the notion that history is indestructible both at the levels of ontology and epistemology. It is this principle of information indestructibility that gives credibility to historical studies, anthropological studies, archeological studies, geomorphological studies and sub-fields that may arise in them such as cultural anthropology and others.

The whole science of environmental transitions, its impact on geomorphological formations and ontological transformation from antiquity to the present is motivated by the *principle of indestructibility of information.* Such information cannot meaningfully be claimed to reside in the mind neither can it be used to define the mind, where the mind is viewed as the immaterial information in the brain. By viewing the mind as an immaterial information, the information is not defined but is used to define the mind in that the mind is the information as seen in either a mind-matter or *BIT-IT* problem in defining the concept of information. If the mind is immaterial information, then how does it acquire the cognitive properties of informing, learning and knowing? There is one important thing that distinguishes information from both matter and energy, and that is information has the property of continual accumulation while matter and energy are infinitely closed under ontological transformations through rearrangements of existing characteristics of matter and energy. Without information, matter and energy are not defined and distinguished. However, without energy information communication is impossible.

2.1.2 *The Conceptual Foundation on* from IT-to-BIT-to-IT

In a search for a general definition of the concept of information before exploring the degree of its content, measurement, properties and uses, it becomes necessary to distinguish between the ontological space and its constituent elements, and the epistemological space and its constituent elements. The relational structure is such that there is a set of *ontological elements* which are the representation of the *IT*, and a set of *epistemological elements* obtained through the *BIT*-process which must be considered in the search process. The *IT* is defined in the ontological space by a natural process while the *BIT* is defined in the epistemological space by some epistemic process. The ontological space with the ontological elements is taken as the *primary category of existence* while the epistemological space with the epistemo- logical elements constitutes the *derived category of existence* obtained by an epis- temic process from the ontological space. The epistemological space is non-existent without the ontological objects entering into active relations with other ontological objects to develop various degrees of knowledge and freedom [R4.7, R4.10, R8.13, R8.16, R8.18, R8.45]. Given the conditions of existence of the ontological space, everything that is in the universal system takes place in the epistemological space through the activities of ontological agents and their acquaintances. The active relationships among the ontological objects are through their acquaintances which may be mutual or non-mutual. The question then is what acquaintances produce from the ontological objects to establish relationships. The acquainting processes generate elements that allow the multiplicity of relationships to be formed. The results of these source-destination connectors among the ontological objects will be broadly classified as information.

The information informs about a presence of objects which may be states, processes and diverse events in the environment of ontological existence. The actualization of these acquaintances takes place in the epistemological space and is recorded in cognition as an *awareness* and subjective *experiences*. What is the awareness and what is experienced? The acquaintance *informs* about a presence or an existence of some ontological object which may or may not be translated into *knowing*. For the ontological objects to inform about their presence they must have some informing attributes. It is the presence of these attributes that must constitute the foundation for defining the general concept of information that contains the phenomenon, qualitative-quantitative dispositions, measurement, properties and uses. On these bases, three questions tend to arise. What informs and who is informed? What is known and who knows it? What is the relational structure of the informing and the knowing processes? The degrees of acquaintance, awareness, informing and knowing provide entry and departure points into conceptualizing the nature of the epistemological space in terms of the nature of assumptions imposed in order to define the parametric structure of the epistemological space over which one is cognitively working. These assumptions will also affect the nature of epis- temic activities and methods of learning and knowing as well as claims and veri- fications of knowledge.

The elements of both the ontological and epistemological spaces present awareness through their sending and receiving signals of their constituent properties that aggregately define their existence. The awareness of the ontological space may be defined by *ontological signals* the collection of which helps to establish a definition of the ontological space through the characteristics of the ontological elements. Similarly, the epistemological space may be defined by *epistemological signals* the collection of which helps to establish a definition of the epistemological space through the awareness of the ontological elements that are transformed into epistemological elements. The collection of ontological signals will constitute the basis to define the *ontological information*. The collection of epistemological signals will constitute the basis to define the *epistemological information*. The information is thus a surrogate of signals received through the source-destination process. The ontological information is the primary category while the epistemological information is derived from the ontological information. The signals, and hence the information in both ontological and epistemological spaces connect the ontological objects and their epistemological derivatives. By logically dividing the concept of information into ontological and epistemological information and logically uniting them through the principle of relational structure of ontological objects, it is here stated and maintained that information is a property of matter without which information is not definable.

As presented, both the ontological and the epistemological spaces are information defined and connected. It is this information connectivity that give rise to methodological nominalism and the rise of a *family of ordinary languages* (FOL) creating communicational signals among human cognitive agents over the epistemological space. It also gives rise to the development of a *family of abstract languages* (FAL) in the epistemological space that may be used to create communication signals among human cognitive agents, among non-human agents such as machines and among cognitive agents and machines. It is through the FOL and FAL that learning and knowing take place. It is through this understanding that learning machines can be constructed in more complex situation. It is useful to keep in mind that both FOL and FAL are representations of the *BIT-process* and that the cognitive agents and non-cognitive agents are representation of the *IT-process*. There is no language and there is no *BIT* in the ontological space. The BIT-process is a property of the ontological space. The ontological information is perfect in terms of exactness and completeness. It is these properties of exactness and completeness of the ontological information that lead to the claim that there is no uncertainty, vagueness, risk or accident in the ontological space. In terms of ontological transformations, the ontological information is also the ontological knowledge in the natural decision-choices that drive the transformations and categorial conversions in the ontological space [R17.15, R17.16, R17.18]. The ontological information is natural, governed by ontological processes and independent of activities in the epistemological space. This does not mean that cognitive agents are independent of ontological processes. In fact, every activity of any ontological element affects the universal environment no matter how small the effect.

Things are different when one critically examines the epistemological space relative to the ontological space which provides identities of ontological varieties from which the identities of epistemological varieties are correctly or incorrectly derived. It is useful to note that corresponding to each ontological or epistemological identity is a defined variety. The epistemological space is a cognitive construct and the epistemological signals are reflections of acquaintances and conceived to constitute the epistemological information. The epistemological signals may pass through noisy channels to generate defective and deceptive epistemological information that will reduce the ability to generate completely worthy knowledge as input into the decision-choice system. The defective component of the epistemological information is the result of quantity limitation of sender-receiver signals from the ontological objects to create *incomplete information structures*, as well as quality limitation of origin-destination signals from ontological objects to create *vague information structures* [R21]. The assumptions of information completeness and information vagueness will define the nature of the epistemological space in which the cognitive agents operate to manage the commands and controls of the decision-choice systems. One epistemic process is to impose an *exact epistemological space* whether the information is complete or incomplete irrespective of the definition of the concept of information imposed to restrict the useful domain of analysis, thought and synthesis. The assumption of an exact and complete information structure defines a *perfect epistemological information space* where there is no uncertainty and risk. This space will be referred to as the *exact non-stochastic epistemological space* [ENES].

The assumption of an exact and incomplete information structure creates a *limited epistemological information space* where there are uncertainties and risks generated by a quantity limitation of information. This space will be referred to as the *exact stochastic epistemological space* [ESES] with *quantitative uncertainty and risk* associated with thinking and the decision-choice system. The paradigm of thought corresponding with the exact epistemological space is the *classical paradigm* which is enhanced with probabilistic reasoning to deal with the limited information component of exactness. The basic foundation of the classical paradigm of thought is the absoluteness of existence of opposites in separable forms without relational continua and unity. This is consistent with the concept of dualism as it has been outlined in Chap. 1 and in [R4.7]. This way of thought development in the created epistemological space rejects contradiction as a valid truth-value, and where the existence of contradiction is used as method of verifying validity of propositions and proof of theorems. This approach to the construction of the epistemological space is to reject important limitations of cognitive agents. The case of exact and complete information produces an epistemological space that is a replica of the ontological space where information is not only perfect but is the same as knowledge. The case of exact and incomplete information is an admission of cognitive limitations through acquaintances, but a lack of admission of the limitation in the FOL and the FAL in the epistemic representations where imprecision and vagueness are assumed away.

The other epistemic process in creating an epistemological space is to allow all the limitations of cognitive agents to define the epistemological space where there is no claim of absoluteness, where conflicts, imperfections, subjectivity and relativity are the fundamental attributes of cognitive existence, and, where vagueness and imprecision are also fundamental attributes of the creation of the FOL and the FAL in all representations and communications. The fundamental attributes of vagueness, imprecisions, conflicts, non-absoluteness and relativity are the result of the existence of opposites in a multiplicity of relations. This is the principle of opposites in relational continuum and unity that is generated by polarity and duality with relational continuum and unity. The result of these attributes is the creation of an *inexact epistemological space* whether the information is complete or incomplete.

The assumption of inexact and complete information structure defines an *imperfect qualitative epistemological information space* with uncertainties and risks due quality limitation. This will be referred to as the *fuzzy epistemological space* (FES) with *fuzzy uncertainty and risk*. The assumption of inexact and incomplete information structure defines *imperfect quantitative-qualitative epistemological information space* with both qualitative and quantitative uncertainties and risks. This will be referred to as the *fuzzy-stochastic* or the *stochastic-fuzzy epistemological space* with *fuzzy-stochastic risk* associated with thinking and the social decision-choice systems. The paradigm of thought that corresponds to the inexact epistemological space is the *fuzzy paradigm* which must be enhanced by probabilistic reasoning to deal with the limited information component of inexactness, as well as by possibilistic thinking to deal with components of contingencies and potentialities of the information and information processes

The fuzzy-stochastic information that defines the fuzzy-stochastic space is further complicated by the possible presence of intentionality of *false information transmission* between sources and destinations of cognitive agents to affect the decision-choice outcomes in social set-ups. The false information that is intentionally transmitted is referred to as the *deceptive information structure*. This defective information structure is made up of two important components of *disinformation and misinformation sub-structures* which are semantic in nature. The deceptive information structure can be weaponized to change the efficiency and effectiveness of the management of the command and control structures of the decision-choice systems. It can also interrupt and change the encoding during the transmission and thus affect the decoding of the information at the destination. The presence of deceptive information introduces an important element of the subjective phenomenon that violates the requirement that DOS (Declarative objective semantic) information be truthful. The deceptive information structure may be used in all communications among cognitive agents over the epistemological space where advantage is sought in any duality and polarity. In this respect, the decoding process may be handled with the tools of the fuzzy paradigm and hence the condition of truthfulness must be dealt with in the fuzzy-non-stochastic space or fuzzy-stochastic space with fuzzy probabilistic reasoning [R4.8, R4.9, R4.13]. Further questions arise as to the relationship between objectivity and truthfulness.

Some important difficulties with knowledge production and the decision-making process arise from the fact that the concept of information is of a diverse nature even within the same specialized discipline. These are difficulties that arise from subjective perception of objective information as one traverses from the ontological space to the epistemological space via an epistemic vehicle. In fact, in the decision-making process, the objective information is not important as input. The most important thing is the subjective perceptive formation of acquaintance of objective information when the concept of information is defined and explicated. It is useful to refer to Fig. 1.2.2 where a complete distinction is made between *objective information* and *subjective information*. Subjective information is a phenomenon of the epistemological space while objective information is the phenomenon of the ontological space. Both subjective and objective information structures have the attributes of *qualitative and quantitative dispositions.*

The concept of information as defined in the epistemological space must relate subjectivity to qualitative and quantitative dispositions, where the objective attribute is restricted in the non-absolute domain in a fuzzy space. It is here that the *family of fuzzy abstract languages* (FOFAL) with fuzzy paradigm of thought composed of its mathematics and logic becomes extremely useful in dealing with innovations in communications and the development of learning-thinking machines under the principle of self-correction in the framework of human learning-thinking systems. The relational linkage between human learning systems and non-human learning systems requires a search for a definition of a general concept of information explicated to fit specific analytical and decision-choice needs. The purpose of this search is to find a common concept of information from which specific sub-definitions may be abstracted with a family of communication and communication channels among diverse ontological objects.

2.2   Seaching for Defining Factors of a General Structure and Definition of the Conceptand Phenomenon of Information

It has been pointed out in Chap. 1 of this monograph that the universe may be conceptualized as an integrated and inseparable system of matter, energy and information, where matter, energy and information mutually define their individual and collective existence. Matter exists as a collection of varieties and categorial varieties in relational continuum and unity; energy also exists as a collection of varieties and categorial varieties in relational continuum and unity; and similarly, information exists as a collection of varieties and categorial varieties in relational continuum and unity. The system is such that energy and information reside in matter where information is a special linkage between matter and energy. Different appearances of matter are forms of the same abstract matter that constitute the general concept of matter, where specific forms are derivable by methodologies of constructionism and reductionism, and where the general matter constitutes the *primary category of existence* such that matter cannot be *created* or *destroyed* but merely be *transformed* into different forms or varieties with identities. Different forms of the disposition of energy called varieties such as chemical, solar, atomic, electric and others are the same form of abstract energy that constitutes the general concept of energy, where specific forms of energy are derivable by methodologies of constructionism and reductionism, and where the general energy constitutes the *primary category of energy* within the energy field such that energy cannot be destroyed or created but be transformed into different forms. However, energy is a *derived category* of existence from matter where each variety of energy can be mapped onto a variety of matter. Similarly, different conceptual dispositions of information are forms called varieties of the same abstract information that constitutes the general concept of information, where specific forms are continually being created by methodologies of nominalism, constructionism and reductionism, and where the general information constitutes the *primary category of existence* such that information cannot be destroyed but is under continual creation and accumulation. It is these properties of continual creation and accumulation of information that separate the phenomenon of information from the phenomena of matter and energy. Both matter and energy cannot be destroyed or created, but information also cannot be destroyed, however, unlike matter and energy, information is under continual creation and accumulation.

An important element of distinction may be noted about information relative to matter and energy. Matter and energy can neither be created nor destroyed but are under continual transformation under the principles of categorial conversion and Philosophical Consciencism of internal arrangements of elements in the universal system, [R17.15, R17.16, R17.18]. Information, however, is under continual creation, storage and accumulation. Thus, Information constitute stock and flow where there is neither stock or flow equilibrium but continual accumulation, while matter and energy are always in stock equilibria with continual internal dynamics and categorial arrangements. It is the definition of a general concept of information that is being sought. The common abstract forms of matter, energy and information constitute the *ontological uniformity of existence*. The conceptual diverse forms of matter, energy and information in the learning-knowing process constitute the *epistemological diversity of existence* that is the work of the epistemic process. The essential analytics of the ontological uniformity of existence is the provision of necessary and sufficient conditions for the development of the theory of categorial conversion of forms of matter and energy under methodological nominalism, constructionism and reductionism, where primary categories of varieties are identified and derived categories of varieties are explained leading to explanatory and prescriptive theories of existence of the *ontological family of varieties* and the *family of categorial varieties* at both the levels of info-statics and info-dynamics over the epistemological space. The methodological constructionism and reductionism allow the linkage of the corroboration-verification process between the primary forms and derived forms through the activities within the information-knowledge duality [R17.15, R17.20, R17.25, R17.35, R17.50].

The organicity of the universal existence is such that the primary category of existence is the *basic matter* from which different forms of matter arise through categorial conversion where the conversions are the results of complex interactive processes of energy and information. But what is information? The rise of different forms of matter from within creates a distribution of different varieties of matter which generates different energy distributions with supporting different distributions of information structures. The concept of information and the information structures must be defined in a manner that allows the establishment of relational continuum and unity of varieties of ontological existence which must then be translated into epistemological existence by an epistemic process. It is useful to keep in mind that ontological existence is the primary category of existence while epistemological existence is the derived category of existence. It is also useful to keep in mind that different forms of matter will not arise by categorial conversion without active interactions of energy and information that reside in the basic matter. It is from this basic matter that we shall pull out the definition of information and define epistemic paths of explications that are useful for specific needs of knowledge areas as they relate to ontological and epistemological decision-choice systems. The energy and information provide the categorial decision-choice processes for the internal qualitative transformation of one form of matter to another form.

How does one know the differences among types of matter and thus place distinctions for recognition? How can the periodic table be constructed and under what conditions? Under what set of conditions does one distinguish a female from a male, a tree from an animal, or gold from a diamond? Consider a message in a source-destination duality of the form where the source indicates that a package of gold has been sent to a specified destination. What set of conditions will establish the truthfulness of the overall content of the message in the source-destination duality? In other words, what is the set of conditions that allows distinctions and similarities to be placed on elements to create varieties? It is from the establishment of an appropriate set of conditions that a general definition of information must be constructed. It is also from this set of conditions that one must understand the concepts of information indestructability in the theory of info-statics and continual accumulation of information in the theory of info-dynamics. There are three complex distributions that are available for analytical work. Every existing form or disappeared form has an information tray of history. This has nothing to do with epistemological record keeping and is independent of individual cognition. However, it has something to do with *ontological record keeping* that offers a pathway to the beginning of times without which our scientific search for other forms of existence in the universe is meaningless and useless. The disappearance of the old does not lead to the disappearance of its information structure. However, the emergence of the new brings into being the supporting information structure that presents new conditions of its existence and the possibility of its disappearance through potential acquaintance. The old information is not destroyed and the new information is added on for a continual increased accumulation of the ontological history which is infinitely large and different from recorded and non-recorded epistemological history. It is also this indestructibility of information that defines the usefulness of research into the discovery of *what there is* and the motivation to find elements of the ontological identity which are hidden from immediate cognition, and places limitation on the human capacity to know, as well as defines the self-correction process of the knowledge-production system over the epistemological space.

Interestingly, it is the same information indestructibility that constitutes the essential motivation in the search for the origins of the universe, and possible existence of other life forms all through the Pharaonic times from ancient to the present. It is the continual accumulation of information that creates insurmountable difficulties and complexity, where illusions of knowledge arise in the possibility space and cognitive imagination in the possible-world space. The general definition of information must be such as to incorporate these essential elements of universal existence in both the ontological and epistemological spaces. Like energy and matter, it must hold an element of uniformity and yet have a property of universal variety.

The search for a general definition of information seems to break down to a number of attributional levels of intentionality at the qualitative nature when involving humans and non-humans. At the level of humans, there is verbal and non-verbal representations and transmissions. The verbal representation and transmission component of the concept of information belongs to the FOL (family of ordinary languages), where the intelligibility and usefulness within the source-destination duality have semantic attributes, broadly defined, given a par-ticular FOL, that affect the behavioral outcomes with the learning-knowing process and the resultant direction of the management of the commands and controls of the human decision-choice system. The non-verbal representation and transmission component of information belongs to the FAL (family of abstract languages), where the intelligibility and usefulness within the source-destination duality have impor-tant qualitative attributes that may or may not include semantic attributes which together affect the behavioral outcomes within the learning-knowing process and the direction of the management of commands and controls of human and non-human decision-choice systems. One thing that stands out clearly as a general definition of information is being sought is that information is active, and in fact the life-blood of all action in the universal existence. This life-blood contains the contents and conditions of transmission.

The general definition of information must capture all these qualitative attributes to ensure the qualitative disposition of information whose presence will affect the quantitative disposition of information and the type of measurement that may be constructed. The question that arises is what this general definition should be. To answer this question, one must find the solution from among all the conceivable concepts of information that engrain qualitative attributes which define the essential elements of the specific concept of information. This should include concepts of human and non-human information structures. To illustrate the problem that emerges from the question, let D be a set of definitions of the concept of infor-mation, then $D = \{D_i | i \in I\}$ where I is an index set of available and potential individual specific concepts of information. Every specific concept is defined by its *qualitative disposition* ($\mathfrak{A}$) and *quantitative disposition* $\mathfrak{B}$ such that one can write a definitional function of the form $D_i = d_i(\mathfrak{A}_i | \mathfrak{B}_i, t)$ such that for any time period and quantity, the specific concept of information depends on the qualitative disposition. The set $D_i = d_i(\mathfrak{A}_i | \mathfrak{B}_i, t)$ must be aggregate to arrive at the general definition of the concept of information that encompasses all specific information concepts that are presently available and may be constructed in the future. This aggregation involves qualitative variables whose values are expressed in the fuzzy space in terms of degrees of importance in contributing to the definition [R4.13]. The *degree of qualitative importance* may be considered as a fuzzy number which may be specified as a fuzzy membership function of the form $\mu_{D_i}(d_i(\mathfrak{A}_i | \mathfrak{B}_i, t)) \in (0, 1)$.

This membership function is increasing from the minimum of zero to the maximum of one.

From the viewpoint of principle of opposites, it is useful to design the *degree of qualitative unimportance* of the input of the definitional function of the form $B_i = d_i(\mathfrak{A}_i|\mathfrak{B}_i, t)$ with a corresponding membership function of the form $\mu_{B_i}(d_i(\mathfrak{A}_i|\mathfrak{B}_i, t)) \in (0, 1)$ that is decreasing from one to zero. The logic of reasoning is based on the principles of duality and polarity with continuum and unity where every input is both qualitatively and quantitatively defined in some relative disposition. The decision-choice action for each input into the construct of the general definition of information is such that the degree of qualitative effect is constrained by the degree of quantitative effect. The general definition of information requires the maximization of qualitative disposition subject to quantitative disposition of the defining variables. The problem is simply a fuzzy optimization problem at each point of time and for each input into the defining function of particular concepts of information. This problem may be stated as a fuzzy decision-choice problem in the form $\Delta_i = (D_i \cap B_i)$ and with a membership function that may be specified as:

$$\mu_{\Delta_i}(\cdot) = \left(\left[\mu_{D_i}(d_i(\mathfrak{A}_i|\mathfrak{B}_i, t)) \right] \wedge \left[\mu_{B_i}(d_i(\mathfrak{A}_i|\mathfrak{B}_i, t)) \right] \right) \in (0, 1) \qquad (2.2.2.1)$$

The fuzzy optimization problem may then be defined in two sequential steps of

$$\mu_{\Delta_i}(\cdot) = \begin{cases} \max_{\mathfrak{A}_i} \left[\mu_{D_i}(d_i(\mathfrak{A}_i|\mathfrak{B}_i, t)) \right] \\ \text{ST.} \quad \left\{ \left[\mu_{D_i}(d_i(\mathfrak{A}_i|\mathfrak{B}_i, t)) \right] - \left[\mu_{B_i}(d_i(\mathfrak{A}_i|\mathfrak{B}_i, t)) \right] \right\} \le 0 \end{cases} \qquad (2.2.2.2)$$

It may be noted that the fuzzy optimization structure fit into an analytical duality with relational continuum and unity under the general principle of opposites in the sense that for every degree of qualitative importance, there is a degree of qualitative unimportance. One may conceptualize the problem-solution process in terms of cost-benefit duality where every cost has a corresponding benefit that gives meaning to the whole decision-choice system and transformation-substitution process. The optimal solution to this qualitative importance problem may be used to form a fuzzy partition in order to form categories and varieties.

The underlying structure of the optimal definition problem of the defining factors of the phenomenon of information and the conceptual problem of the definition of information as represented by the Eq. (2.2.2.2) is a *signal disposition* that must be related to qualitative and quantitative dispositions of elements. This problem is part of all the epistemological problems of knowing which include the phenomenon of information. Any concept of knowing implies a distinction that involves a distribution of varieties that allows categories of varieties to be created in static and dynamic domains with neutrality of time. Such *varieties* will be shown to emerge from *characteristics* which present continual signals in universal existence. This approach to the phenomenon of information leads to a non-standard and general information theory. Here an argument is put forward that the concepts and measurements of anticipation, uncertainty and surprise cannot be used to define information since such an approach involves circularity of reasoning that offers us no way out of critical cognition. Similarly, the concepts and measurements of possibility and probability offer us no way out in understanding the structure of the phenomenon of information and its definition.

To proceed in the development of non-standard information, it is useful to provide an epistemic reflection of the main ideas of the conceptual definition of standard information theory which involves the definition and structure of information. The theory of non-classical information that is being developed here has become necessary as the result of the problems encountered in all application to which the concept of information is used. The concepts of information and knowledge were introduced in Sect. 1.1 without explicit definitions. It seems that, in most theoretical and empirical works in exact and inexact sciences, information and knowledge acquire interchangeability without specifying the conditions of their substitutability. In this way knowledge is information and information is knowledge. The information-knowledge equality only holds over the ontological space in which case there is no uncertainty, ignorance, surprise and risk. It is useful to state what the concept of information is not by keeping in mind the process-path to knowledge discovery.

The first thing to keep in mind is the phenomenon that may lead to the naming and creation of a concept in terms of vocabulary within a particular language through the principle of acquaintance. The next step is to clean the concept from what it is not and set the framework for its definition and possible explication. Given the definition and explication, it becomes useful to examine the content of the concept (semantic containment, quality), where the content of the concept is subjected to conditions of measurement and unit of measurement (quantity). In general, one cannot meaningfully discuss the content of an unknown concept as well as measure the content of an unknown concept. This is the quality-quantity problem of all concepts in the general knowledge-production system when they are given definitions and explications for clarity, variety and distinction. The steps of the non-standard approach as distinguished from the standard approach in Fig. 1.1 may be presented in a cognitive geometry as in Fig. 2.1. It is useful to observe that in the establishment of the analytical process of the classical information structure and the theoretical development, the phenomenon of information is missing. In this respect, the classical theory of information deals substantially with the quantity side of information. Even the semantic approach to information is also guilty of this emphasis on quantitative disposition as it relates to subjective probability of false-truth acceptance of statements. The question that may be asked is: in what process can the concept of information incorporate both quality and quantity elements of the phenomenon of information in the classical approach and utilize the epistemic machine of the classical paradigm of thought? An important distinguishing factor of the non-classical approach to defining the concept and phenomenon of information is that information is accepted as a property of matter and

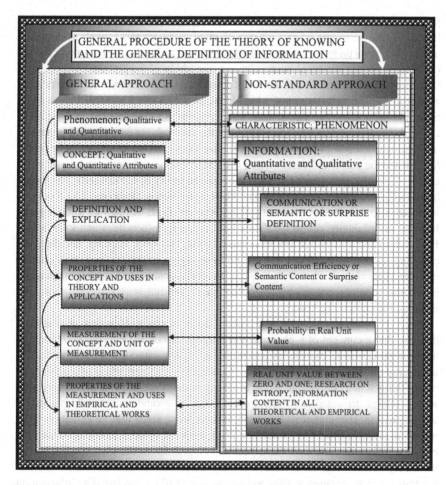

Fig. 2.1 A cognitive geometry the process of knowing and information structure

energy where the three exist in relational continuum and unity with matter constituting the primary category of existence and energy and information constituting the derived categories of existence.

2.3 The Theory of Knowing and Information Definition a the Non-standard Approach

The theory of knowing is directed to answer a class of fundamental questions. (1) What does the knower know and how well is this known (content)? (2) What is the motivation to know (curiosity and survival)? (3) What are the methods and procedures of knowing (laws of thought, thinking and reasoning)? (4) How does the knower know he or she knows what is claimed to be known as well as convince others that something is known (verification, truthfulness and communication)? (5) What is the usefulness of what is known (the utility of the known)? This class of fundamental questions of the theory of knowing is different but intimately connected to the theory of learning which must answer a class of fundamental questions. (1) What does the learner learn (the content)? (2) What is the motivation to learn (curiosity and survival)? (3) What are the methods, techniques and processes of learning (methods and techniques)? (4) How does the learner know that he or she has learned the content of what is to be learned (content verification and assurance)? (5) How useful is that which has been learned (the utility of the learned). The theory of knowing is the foundation of all knowledge-production processes. The theory of learning is to ensure the continuity of the known. Both theories of knowing and learning are mutually connected. They both relate to information and knowledge. Information and knowledge are phenomena, while knowing, learning, deciding and choosing are non-communication and non-messaging processes involving content creation. Teaching, on the other hand, is a communication and messaging process involving content transmissions that are intimately connected in the information-knowledge space.

Every phenomenon is defined in quality-quantity space. However, it is always the case that a concept of a phenomenon may be introduced into the knowing and learning processes to capture its *qualitative existence* without a known process of measuring the qualitative content. Here, the semantic containment is of a qualitative nature defined in a penumbral region of clarity in the space of subjective disposition which relates to qualitative disposition. The phenomenon of information is unspecified in the classical information structure. Since the information phenomenon is unspecified one is not sure of the concept of information on which the classical information theories have been developed. From the general structure of the theory of knowing, the definition of the concept of information must not include the transmission, communication and measurement of its content. The information structures defined by using the concepts of transmission, semantics and communication and content measurements belong to a class that is here referred to as the *classical information structure* otherwise, it will be referred to as the *non-classical information structure*. In general, there are theories of information coding, transmission, communication, measurement and others which are simply sub-theories of the general theory of information which is intended to examine the phenomenon of information and information-production. It is not different from the theory of knowing which is intended to examine the phenomenon of knowledge and knowledge-production. The theories developed around the classical information structure using the classical paradigm of thought belong to a class which is, here, referred to as the classical information theory. The theories developed around the non-classical information structure using the non-classical paradigm of thought belong to a class which is, here, referred to as the non-classical information theory.

The definition of information, however, includes things that inform other elements in the universal space. The concept of information in the process of knowing must be viewed from two interdependent components of *characteristics of objects* that express the morphology of ontological elements, and *relationships among objects* that express the structure and form of epistemological elements. The characteristics of objects are qualitatively defined to bring into view the categorial differences in the ontological space. On one hand, information constitutes the general set of the overall properties of objects, states and processes, and hence presents the set of identities of objects in addition to their behaviors as managed by internal command and control decisions at the state of ontology. This is the *ontological information structure*. On the other hand, information is a set of relationships among objects as they pass through states and processes by means of sender-recipient moduli which offer necessary and sufficient possible conditions for the management of commands and controls by socio-natural objects under continual transformations, and in particular by cognitive agents in the epistemological space. This is the epistemological information structure. The former is *objective* in the sense that it exists independently from the awareness of other ontological entities. These entities constitute the *ontological identities* which exist only in the ontological space. These identities provide *ontological varieties*. The collection of similar varieties constitutes a category in the ontological space. The ontological categories are an infinite family of categories in the universal system. The latter is *subjective* in that the relationships and their types that are formed require the awareness of other ontological entities. It is a set of signal transmissions with mutual give-and-take subjective activities which exist in the space of knowing to produce epistemological entities. The subjective information reveals derived elements as *epistemological identities* to the cognitive agents.

The revealed identities constitute derived epistemological elements which reside in epistemological space and constitute the set of epistemological varieties. The collection of the same revealed elemental identities constitutes epistemological categories as defined by qualitative signals. The epistemological categories are a finite family of derived categories in the space of knowing. Information is the connecting link between the ontological and epistemological spaces, and hence is defined in terms of *objective-subjective* duality as relationally viewed in term of characteristics and relationships that characterize the universal system of objects, states, processes and events. The universal system of objects, states, processes and events is established within the matter-energy polarity that exists in relational continuum with information as the connector.

Chapter 3
The Theory of Info-Statics: An Epistemic Unity in Defining Information from Matter and Energy

The concepts of matter, energy and information were introduced in Chap. 1 and enhanced in Chap. 2 as a search for factors relevant for a general definition of the concept of information was sought to allow one to place an epistemic distinction among them. At the level of cognition, there are definitions for the phenomena of matter and energy that allow one to place distinctions. It has been pointed out that both energy and information reside in matter and that matter constitutes the primary category of universal existence while energy and information are derived categories of existence. The three reside in an inseparable unit where matter is linked to energy by information. In terms of knowing, information constitutes the primary category and matter and energy constitute derived categories. In this epistemic structure and basic existence, information is not linked to matter by energy. Information must help to establish differences and similarities between matter and energy, and among varieties of matter as well as varieties of energy, where varieties of information are identified by adjectival constraints. There are as many varieties of information types as there are varieties of ontological existence in the actual and potential spaces. To understand the logical connectivity of matter, energy and information as defining the basic form of universal existence as one abstract element existing in three and as three elements existing in one, it is useful to have one *general definition of information* that allows one to keep track of the phenomenon of information and its possible contents.

Let us accept the abstract definitions of matter and energy and the corresponding set of measures that have been designed so far, and then ask the question as to how can a general abstract definition be constructed for information that will ensure the fulfillment of its role to place a distinction between matter and energy and serve as a *decision-choice connector* of matter to energy and energy to matter. What form must a general definition of information take in order to also serve as a *cognitive connector* between the ontological space and the epistemological space such that a logical distinction can be established to define an *information-knowledge duality*? At the level of ontology, this general definition must establish an objective

© Springer International Publishing AG 2018
K.K. Dompere, *The Theory of Info-Statics: Conceptual Foundations of Information and Knowledge*, Studies in Systems, Decision and Control 112, DOI 10.1007/978-3-319-61639-1_3

relational structure between matter and energy, while at the level of epistemology, it must make explicit the relational structure of the phenomenon of the *objective-subjective duality* in continuum and unity. Again, at the level of ontology, the general definition that is sought must establish an objective relational structure among forms of matter on one hand, and forms of energy on the other hand, while at the level of epistemology, it must make explicit the relational structures of conditions of forms of the phenomenon of existence in the space of *objective-subjective duality* in continuum and unity. The general definition of information must also establish the phenomenon of information and its contents at both ontological and epistemological spaces. It has been argued and explained in a number of places that the ontological space is the universal *primary identity* while the epistemological space is a *derived identity* [R4.13, R17.15] on the basis of cognition. In this respect, the general definition of information must establish a primary identity of information in the ontological space and a derived identity of information in the epistemological space. The decision-choice connector relates to the concept of *info-dynamics* and continual natural transformation while the cognitive connector relates to the concept of a *self-corrective knowledge-production system* with continual social transformation by cognitive agents through a number of decision-choice systems.

3.1   The Concepts of Ontological Variety and Category

In Chaps. 1 and 2, the concepts of matter, energy and information were introduced to help keep track of factors that will be useful in the search for a general definition of information. What are the relational structures among matter, energy and information that will provide an epistemic pathway to establish the factors that will be useful to abstract a general definition of information? At the ontological space the materialist position is asserted that matter has independent existence that is endowed with internal self-motion. Matter gives rise to energy, information is carried by energy and the differences and similarities between matter and energy are established by information. This matter-energy-information relational structure reveals two important postulates that bring into focus the needed factors that must define the flexible boundaries for the establishing of a general definition of information. They are the existence of conditions of *ontological variety* and *ontological category* at the level of ontology. The ontological variety and ontological category form the basic structure of the universal existence of matter and energy. It is from the conditions of ontological variety and category that the general definition must be abstracted as the primary definition and from which other definitions may emerge as derived. The ontological variety and category express conditions of difference, similarity and objectivity of existence which must constitute the factors in defining information.

3.1.1 Conditions of Ontological Variety and Category

The conditions of similarity and difference are due to inherent attributes that endow matter and energy with varieties, categories and inter-categorial relation of qualitative and quantitative differences and similarities. These attributes are the *ontological characteristics* that present themselves in qualitative and quantitative dispositions with powers of internal self-motions. The internal self-motion of matter leads to the internal self-destruction and self-creation of *ontological varieties* without any destruction of matter. Here, we have qualitative varieties and quantitative varieties. Every ontological variety is completely defined by a *finite characteristic set*. The universe of matter and energy is established by a *universal characteristic set* where by the materialist conception of existence of varieties is acknowledged. It is from this universal characteristic set that a characteristic set of any ontological variety is derived as a proper subset. Every universal object is a variety that belongs to a group of the same variety. The group of the same variety, at the level of ontology, is called the *ontological category* and each variety in a group is called a *categorial variety* (for example every tree constitutes a variety and is identified by a finite characteristic set while all trees will constitute an ontological category). It is from the acquaintance of these ontological characteristics that ontological objects become informed of their environment and the ontological objects that the environment contains. Given the characteristic sets and subsets, it is appropriate to view the ontological objects as intermediaries in the ontological transformation-substitution process which presents the possibility to maintain a logical structure between the ontological states with continual dynamic processes of variety and matter-energy information structure. It is here that the concept of *info-dynamics* at the level of ontology becomes analytically useful in the study of conditions of destruction and conversions of varieties and categorial varieties. The phenomenon and the concept of info-dynamics will be discussed after a general definition of information is given. This discussion will take place in a separate volume under a title of the theory of info-dynamics. The development of the theory of info-dynamics is not possible without the development of the theory of *info-statics* which involves the condition of a general information definition and relevant elements to establish identity within the space of differences and similarities. The continual dynamic process is about qualitative transformations in terms of *categorial conversions* of ontological objects to ensure continual destruction and creation of varieties to maintain the ontological actions in the *actual-potential polarity* through the internal mechanisms of objects in the ontological space, where matter and energy are preserved by a substitution-transformation process through the dynamics of categorial variety.

The dynamics of categorial variety is the existence of actual-potential differentiations of ontological objects into the expansion of the cardinality of ontological varieties, where old varieties may be destroyed by transformation and new varieties may be created by substitution, or the old varieties may go through further differentiations to maintain, expand or reduce the number of ontological varieties.

Alternatively, new varieties may be transformed from some aspects of existing varieties to expand the cardinality of categorial variety. One must relate these reflections to the idea that ontological objects exist as different varieties of the same substance of matter and by the postulate of the residency of energy in matter, there is a derived postulate that matter is a plenum of forces that endow it with the power of self-motion for internal self-transformations of old varieties into new varieties in a never-ending substitution-transformation process in the actual-potential polarity [R17.15, R17.16, R4.13]. It is the existence of the set of differential ontological characteristics that places categorial distinctions and creates *categorial differences* in the ontological space. Each ontological characteristic sends an *ontological signal* to indicate its presence. Corresponding to the set of ontological characteristics is a set of ontological signals. It is the signals generated by the characteristics from the ontological objects that establish ontological information to reveal ontological identities and corresponding varieties.

Ontological information is a set of ontological signals that reveals the nature of ontological characteristics, and places distinctions and similarities on ontological objects and categories to create categorial differences and varieties. The signal are manufactured and encoded within matter with the use of its internal energy and organization, and then transmitted democratically as messages to all ontological objects. It must be understood that every ontological object is both a producer and a receiver of massages. The ability and capacity to decode the message for understanding demands special internal organization of ontological objects which have cognitive capacity or not. It is here that the problem of communication among machines arises and acquires scientific interests. The problem involves finding the set of internal organization of machines that can internally encode its own signals and decode signals from other machines. In general, every signal is in a continuum and is transmitted in the same manner that is in analogue mode. Digital and discretization are attempts in the cognitive process to understand the information continuum through actions of focusing specificity for sharpness and amplification. At the level of ontology, the signals as messages of information are the surrogate representations of the ontological characteristics which are the content of information. Information is thus defined by characteristics of objects that reveal the nature of their categorial varieties through the characteristic signals. Each characteristic sends one and only one signal as its surrogate representation to indicate its presence that creates capacity for acquaintances by other ontological objects. Generally, therefore, there are two important factors on the basis of which information can acquire a general definition and measurements. These important factors are *characteristic disposition* and *signal disposition*, where behind every *signal disposition* there is a *characteristic disposition* in all definitional forms of information without which the phenomenon of information is vacuous. The use of characteristic disposition and signal disposition as defining elements of information allows information to be naturally connected to matter and energy as organic categories in the ontological space. So what is characteristic disposition and what is signal disposition? These concepts must be defined and explicated.

Definition 3.1.1.1: Characteristic Disposition
A characteristic disposition is a set of attributes that places distinctions and similarities among ontological objects to create ontological varieties and categories in terms of qualitative distinction within which quantitative disposition may be established.

Definition 3.1.1.2: Signal Disposition
The signal disposition is a collection of codes of elements in the characteristic disposition that are transmitted to all categorial objects in order to indicate the presence of an ontological object.

Definition 3.1.1.3: General Information Definition (GID)
Information is a combination of characteristic and signal dispositions that present qualitative and quantitative dispositions. Information is characteristic-signal disposition in quality-quantity space. It is a property of ontological objects (matter) which can be transmitted by energy among ontological objects. Its essential properties are indestructibility and continual accumulation in relation to identification of ontological varieties and categories in the universal system.

Note
The definition of information offered here is the most general that one can think of. This definition is the primary category of the definition of information. All other definitions are categorial derivatives from this primary category. Underlying every signal is a characteristic, where every signal is an energy process which carries the encoded message of the particular characteristic to reveal the identity of a particular variety. The sets of characteristics and signals are finite for each ontological object that reveals its variety and possible category of belonging. Every *event* irrespective of how it is defined is a variety of a future happening as revealed by a *finite characteristic set* which projects a *finite signal set* that is in the past-present-future framework of knowing. Every characteristic is directly or indirectly traceable to matter. The concept of variety presents *inter-categorial difference* and *intra-categorial difference*. The inter-categorial difference is qualitative in nature which defines differences among categories while the intra-categorial difference is quantitative in nature which defines differences within categories. The distribution of inter-categorial differences is defined to distinguish ontological objects in varieties and to place them in respective ontological categories. The distribution of intra-categorial difference is established within categories to reveal quantitative differences such as small and big. The general definition of information establishes the *content of information as a characteristic set* of a variety and the *quantity of information is the number of the characteristics*. In this respect, a signal set is the surrogate of the characteristics of the variety, and hence the measurement of the number of signals *in the signal disposition* of a variety is an indirect measure of the characteristics. It is here that the discussions on channels of transmission, capacity and noise of a channel acquire important analytical demand in knowing. The non-standard information theory that is being presented is about what information is from which other relevant scientific and philosophical questions may be raised and answered.

3.1.2 Reflections on the Information-Defining Factors

The factors which have been abstracted to assist in the development of a definition of general information as categorial-variety driven from the relational continuum and unity of matter and energy are presented as an epistemic geometry in Fig. 3.1. Here, two concepts of information are intended. They are *ontological information* and *epistemological information*. This chapter deals with ontological information as the primary category of the phenomenon of information. It presents the ontological identity in terms of *what there is*, the actual, and *what would be*, the potential in varieties and categories through identification, clustering and category-formation. The epistemological information is a derivative from the ontological information whose foundational definition and phenomenon will provide an important basis for the definition and phenomenon of epistemological information. It will become clear that the conditions for the existence of the concept and phenomenon of ontological information is not different from those of epistemological information. The epistemological information is a derivative by the methodological constructionism while the ontological information is the primary by methodological reductionism. Thus, the non-standard definition of the concept and phenomenon of information will be shown to incorporate different definitional forms of the concept of information and the phenomenon of information. One must keep in mind that the general definition of the concept of information makes information the property of matter through ontological characteristics and formation of varieties.

From the reflection of the necessary conditions of the factors, a message is a signal disposition that contains the information the content of which is the characteristic disposition. The signal disposition is a surrogate of the characteristic disposition as well as a derived category of transmission of existence of varieties. The characteristic disposition is the primary category of transmission that relates to

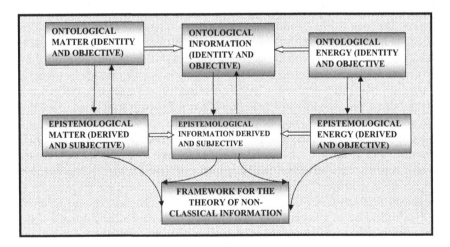

Fig. 3.1 Epistemic geometry of necessary conditions for non-classical information structure

the real definition of the existence of varieties and the corresponding identities. There is no message without a signal disposition and there is no signal disposition without a characteristic disposition. The messaging system is the set of transmissions among ontological objects under the principle of acquaintances through either direct or indirect observations. The messaging system is also the communication system that is consistent with the general information definition as a characteristic-signal disposition. The power of this definition is its ability to relate messages to varieties and identities without which messages are indistinguishable and meaningless. The concern in this definitional process is the identity of the message and not the size of the message that is of concern. The size of the message is always full and clear at the level of ontology.

3.1.3   Information as an Energy-Process

From the beginning of the development of theoretical foundations of informing, learning and knowing contained in this monograph, it was argued that the essential universal aggregates on which the concept of the universe revolves are *matter* and *energy* with *information* as the connector of matter and energy to create relational continuum and unity of ontological objects. In fact, energy is indistinguishable and connected to matter without information. Similarly, categorial varieties of the organic variables of matter and energy are indistinguishable without the distribution of characteristic-signal dispositions. Given the two universal aggregates of matter and energy, the general definition of the concept of information is derived from matter in terms of characteristics which present varieties and categorial varieties in similarities and differences which then fix the contents and phenomenon of information. The matter in collaboration with energy in terms of signals presents motions of the characteristics to fix the possibility for measuring the content of information under appropriate conditions. The characteristics and their signals fix the role of matter in defining the concept of information and establishing the phenomenon of information. An important question, then, arises as to what role energy plays in the concept and phenomenon of information. Here, arises an important relational complexity in informing, learning and knowing about matter, energy and information. The complexity of the relational interplay among matter, energy and information rests on the fact that matter and energy are unidentifiable without information which in turn has no existence without matter and energy. The absence of information, in terms of a characteristic-signal disposition makes it practically impossible to access a particular variety of matter and needed variety of energy and without matter, information does not exist and without energy information cannot be transmitted.

What seems to be the case in this relationally triangular complexity is that without the existence of matter, information is not definable or transmitted. Information about categorial varieties in all forms of existence lives in a static state without energy which is a derivative of matter as the primary category of existence.

Information, defined as a characteristic-signal disposition, cannot play the role as a connector for informing ontological entities without being propelled by the internal energy of matter. Just as energy is a form of a material process, so also is information in a material process induced by the actual and potential energies that create varieties of transmissions of the elements of the signal disposition to present the elements of the characteristic disposition within the actual-potential polarity. In this complex triangular relation, the qualitative and quantitative characteristics of matter present the content of information while the qualitative and quantitative signals serve as surrogates of categorial varieties and are distributed by energy which is then linked to matter by information at the ontological level.

At the level of categorial conversions and ontological decision-choice processes, it is the characteristics that are transformed as the signal disposition to inform and induce conversion though the internal energy of matter. It is here again, that one encounters the creative works of socio-natural processes with quality, quantity and information of matter and energy with neutrality of time. The signal disposition defines a specific threshold of dynamic quantitative disposition beyond which the characteristic disposition assets a qualitative disposition for informing and decision-choice action. The point here, is that information is associated with *identification* and the nature of *creation of categorial varieties of matter*, as well as transmitting the information among ontological objects, where such transmission and decision-choice activities are induced by energy in dualistic actions within the actual-potential polarities. In this respect, the *general information theory* that is being advanced answers the question: what is information; what is transmitted; how is it transmitted and what is measured? *Information is a characteristic-signal disposition among ontological objects. The signal disposition is what is transmitted. The characteristic disposition is what is measured.* The transmission process from an ontological object to other ontological objects is noiseless and may be illustrated in a cognitive geometry as in Fig. 3.2.

In the discussions on the concept of categorial variety, qualitative and quantitative dispositions were introduced and related to the characteristic-signal disposition of ontological objects. It is conceptually important to note that by characteristic-signal-disposition of the phenomenon of information, it is the concept of qualitative disposition that locks in the identity and distinction of informing and knowing in inter-categorial varieties where the concept of qualitative ordering subjectively defines and entails substantial disagreements over the epistemological space. It is the concept of quantitative disposition that forms the basis of informing and knowing in intra-categorial varieties where the concept of ordering within categories is quantitatively induced by the distribution of characteristic-signal dispositions. In the triangular relation of matter, energy and information, the following organic abstractions may be made. The phenomenon of matter concerns the problems of *existence* and *transformations* in the ontological space. The phenomenon of energy is concerned with the problems of *power* and *work* in the ontological space. The phenomenon of information concerns the problems of *knowledge* and *decision* in the ontological space in relation to identification problems and command-control problems.

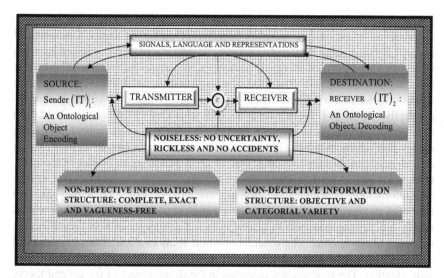

Fig. 3.2 An epistemic geometry of relational structure between transmitter-receiver duality and source-destination duality under conditions of non-standard information in the ontological space

Given the matter-energy-information triangular relation, both energy and information problems are seen as involving the dynamics of qualitative and quantitative dispositions of matter and the implementations of the solutions to the command-control-decision problems with knowledge as input. It may be understood that the mode of transmission and the form of energy required for the work of transmission will depend on the nature of the categorial variety. It is here, that the study and understanding of categorial varieties of energy from varieties of matter become analytically useful. Some ontological information is carried between ontological objects of a source-receiving process by sound energy, and some information is carried by light energy that may be related to the distribution of all other categorial varieties of energy such as thermal, mechanical, electrical, magnetic, chemical, elastic, nuclear, gravitational and other possible ones. The response of the source-destination process and source-receiving process may be carried out by different energy modes.

3.2 The Set-Theoretic and Algebraic Structure of the Definition of Ontological Information

From the discussions in Sect. 3.1, it is useful to bring into focus some working definitions in set-theoretic and algebraic framework which will allow one to see how the factors were abstracted in constructing the general definition of information from the ontological space. It is also important to conceptualize the relational complexity of matter, energy and information in continuum and unity and to answer a number of questions. Is energy a property of matter or matter a property of energy? Is information a property of matter or matter a property of information? Is information a property of energy or is energy a property of information? Can energy exist without matter and what kind of energy will it be? Can information exist without matter and what kind will it be? Can energy and information exist independently of matter? If the answer is no, then what kind of relationships hold them together? If the answer is yes, then how can they exist outside matter? Do matter, energy and information exist in relational continuum and unity or do they exist in discreteness and disunity? Another way of conceptualizing the series of the questions is to ask the question regarding which one of the existence of matter, energy and information constitutes the primary category of existence if they exist in relational continuum and unity. Another question to contemplate is simply if matter, energy and information exist in discreteness and disunity then how they affect the existence of one another. The general definition of information relates to matter and energy as viewed in terms of sets of characteristics and signals to which we now turn attention. The fundamental postulate in the development of the general theory of information is that information is an inseparable property of matter and by logical extension is also inseparable property of energy. The general theory of information is made up of two sub-theories of the theory of info-statics and the theory of info dynamics as has been stated in this monograph which is devoted to the theory of info-statics.

3.2.1 Abstracting the Universal Object and Characteristic Sets (Characteristic Disposition) from Matter

As it has been argued, the universe is composed of matter which constitutes the primary category of existence from which all other existence are pure derivatives. It has also been argued that this primary matter is composed of varieties and categorial varieties from which the general concept of information as a characteristic-signal disposition may be derived. The first task then is to deal with the nature of the ontological objects in universal existence.

Definition 3.2.1.1: Universal Object Set
The universe is composed of the collection of all objects, states, processes and events that are exhaustive, complete and infinite at any time point, referred to as the *universal set of ITS or ontological objects.* If ω is the generic elemental representation of *its* or ontological objects in the universal system then the collection of all ω constitutes the global unity and is simply the *universal set of ontological objects or its,* Ω at any time, written as: $\Omega = \{(\omega_1, \omega_2 \cdots \omega_\ell \cdots) \mid \ell \in \mathbb{L}^\infty\}$, where $\mathbb{L}^\infty$ is an infinite index set of ω which is also an index set of *ITS* where the *ITS* are also the ontological varieties.

The elements in the universal object set are divided into cognitive objects that have awareness capabilities and non-cognitive objects that have no awareness capabilities. Both cognitive and non-cognitive objects reside in the universal unity. Here the awareness means that an ontological object can receive signals and respond to them according to will, while lack of awareness means that an onto-logical object can receive signals but cannot respond according to will. The cog-nitive objects are *info-active* while the non-cognitive objects are *non-info-active*. The entities in the universal object set composed of objects; states, processes and events are infinite in number. Their existence is objective and defined by *objective information* which is defined by characteristic-signal disposition. Their awareness is subjective and defined by *subjective information* as decoding of the signals. The ontological identities and their existence are revealed by a universal characteristic set in terms of attributes.

Definition 3.2.1.2: Universal Characteristic Set and Characteristic Disposition
If x is the generic representation of *attributes* on the basis of which the ontological objects called *its*, in the universal object set are naturally formed, defined, identified and separated, then the collection of all x constitutes the *total ontological attribute space* called the *universal characteristic set*, $\mathbb{X}$, which may be written as: $\mathbb{X} = \{(x_1, x_2 \ldots x_j \ldots) | \Omega,$ and $j \in \mathbb{J}^\infty\}$, where $\mathbb{J}^\infty$ is the infinite index set of all attributes at any given time point. The characteristic set is the characteristic disposition.

The set Ω defines all *ITS* which are the ontological objects in the universe while, the set, $\mathbb{X}$ defines all *objective qualitative and/or quantitative characteristics* in terms of attributes associated with elements in Ω independently of the awareness of any of the objects in the universal object space. In other words, the existence of any ontological object (an *IT*) does not depend on the objective or subjective awareness of other ontological objects (the *ITS*). Every ontological *IT* exist as a variety. All cognitive and non-cognitive *ITS* or the ontological objects, that which are there, belong to the universal object set Ω which is infinitely closed under category-formation at any cosmological or cosmic time. The set $\mathbb{X}$ is also infinitely closed under attribute collection but changes with transformation and difference. Relationally, there are objects defined by characteristics which present a factor for defining objective information as a General Information Definition in the ontological space. A relational structure between the *universal objet set* and the *universal characteristic set* in abstracting the factors through a cognitive path in defining the general concept of information is shown as an epistemic geometry in Fig. 3.3.

The universal object set begins with matter whose differential existence in terms of identities is obtained by the characteristic identification function **I** that maps the universal object set, Ω onto the universal characteristic set, $\mathbb{X}$. The cognitive path requires the methods of characteristic analytics under both methodological con-structionism and reductionism with association function, clustering modules and processing function where for every transformational function there is an inverse function. The cognitive path is intended to solve the *identification problem* in knowledge production through the phenomenon of information in a sequence of

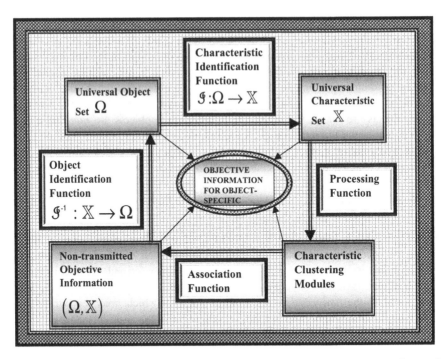

Fig. 3.3 The cognitive path for defining objective information, where **I**, maps the attributes of each *it* in the universal object set into the universal characteristic set and $\mathbf{I}^{-1}$ is an inverse function for object (it) identification

information, knowing, learning and teaching. The characteristics constitute the basic foundation for defining the content and phenomenon of information on the basis of which one can examine the content and the phenomenon of knowledge and knowledge-production in all areas of socio-natural decision-choices and transformations. The characteristic disposition is a set of attributes under the conditions of qualitative-quantitative dispositions.

The elements of any characteristic set are under qualitative and quantitative self-transformation and self-mutation within any relevant category, while every category is under the dynamics of inter-categorial conversion and intra-categorial conversion induced by the internal energy. The point here is that there are continual self-creation of *varieties* in the universal system and that these varieties are identifiable by the corresponding characteristic set as the necessary condition for informing and knowing. Another way to look at this is to see the universe as infinitely closed under continual transformation, where categorial conversion defines the *necessary conditions* of transformation and philosophical Consciencism defines the *sufficient conditions* for the continual transformation, where old varieties may be destroyed and new varieties may be created. The destruction of old varieties does not destroy the history of their existence and the bringing into being of new varieties updates the accumulated history of the existence of ontological objects. The creation of characteristics and categorial varieties are never-ending to maintain the permanency of the universe in an infinite existence and infinite space. The permanency of the universal existence with continual transformation implies *continual accumulation* which in turn implies stock and flow *disequilibria* of information.

3.2.2   Abstracting the Universal Signal Set from the Universal Characteristic Set

The previous section links the concept of categorial variety to the concept of the universal characteristic set to present the distribution of the ontological object set and the identification problem in general. It has to be pointed out that the information phenomenon is actually the identification problem in all existence with or without awareness. The concept of awareness is also the concept of acquaintance and capacity to differentiate by ontological objects in the quantity-quality duality as seen in the actual-potential transformation phenomenon. It is here that the primary matter is linked to the composition of varieties and categorial varieties of matter from which the general concept of information as a characteristic-signal disposition may be derived. The solution to the first task which deals with the relational structure of the universal object set and the universal characteristic set as a way of abstracting a general definition of information toward solving the identification problem is discussed. The collateral argument suggests that the universal characteristic set resides in matter and that this characteristic set has independent existence. Since it resides in matter it is either a direct or indirect property of matter.

Every element in the universal object set has independent existence and the awareness of the existence of other ontological objects. It is the awareness of the composing elements of the characteristic sets that creates epistemological varieties and categories of varieties. The process of awareness is the process of finding a solution to the identification problem. The awareness and the solution to the identification problem of varieties are signal communications that are generated from the elements of the universal characteristic set. Corresponding to the universal characteristic set is the *universal signal set,* $\mathbb{S}$ where each characteristic $x \in \mathbb{X}$ has one and only one signal $s \in \mathbb{S}$.

Definition 3.2.2.1: Universal Attribute Signal Set and Signal Disposition
If s is a generic representation of attribute signals on the basis of which the elements in the universal characteristic set $\mathbb{X}$ are naturally sent from source $\omega \in \Omega$ for identification, then the collection of all such attribute signals from $\mathbb{X}$ is the *universal attribute signal set* $\mathbb{S}$ from the universal object set Ω that may be written as:

$$\mathbb{S} = \left\{ (s_1, s_2, \cdots, s_j, \cdots) | \Omega, \text{ and } \mathbb{X}, \text{ where } j \in \mathbb{J}^\infty \right\},$$

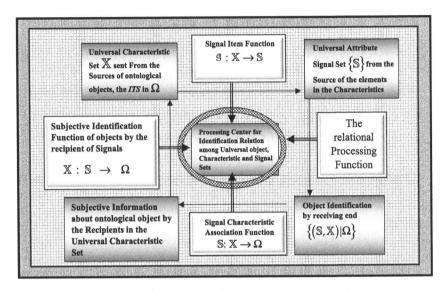

Fig. 3.4 The cognitive path for relating universal signal set to the universal characteristic set to produce characteristic-signal disposition in defining the content and phenomenon of general information as seen in source-destination duality where there is a signal-item function **G** which allows a transmission channels to be created among the elements in the universal characteristic set to creating conditions of objective-subjective duality and quality-quantity duality in the decision-choice system in the command-control dynamics of the actual-potential polarity

where $\mathbb{J}^{\infty}$ is an infinite index set of attribute signals from the source. The attribute signal set is the *signal disposition*.

The elements of the attribute signal set pass through *cognitive filters*, become processed and transformed into a *perception characteristics set* that establishes a *set of information relations* among ontological objects under conditions of acquaintance. Thus, the universal attribute signal set $\mathbb{S}$ defines conditions of *subjective information* which is here defined as *relation-based information* of the elements in $\omega_\ell \in \Omega$ from the universal characteristic set $\mathbb{X}$. Corresponding to the universal object set is the set of phenomena Φ with a generic element $\phi_\ell \in \Phi$ for every $\omega_\ell \in \Omega$. The relational structure of the universal characteristic set and the universal attribute signal set is shown as an epistemic geometry in Fig. 3.4 as creating the conditions of objective-subjective information within the quantitative-qualitative dispositions. The structure is such that for every ontological characteristic there is one and only one signal. In this respect, the size of the infinite set of characteristics is the same as the size of the infinite set of the universal attribute signal set, which technically will be greater than the size of the infinite set of the universal object set. In other words, $(\#\mathbb{X} = \infty^{\infty} = \#\mathbb{S})$ but $> \#\Omega = \infty$ and where (∞^{∞}) is infinity of infinity. The universal characteristic set and the universal signal set provide the necessary and sufficient conditions for establishing variety, categorial varieties and relational structures for *inter-categorial* and *intra-categorial* differences. The signal disposition is a set of signal attributes under the conditions of qualitative-quantitative disposition.

3.2.3   Set-Theoretic Abstractions of the General Information Definition (GID) and Its Content and Phenomenon

The conditions and needed factors for providing a general information definition (GID) as the primary category of the content and the phenomenon of information has been abstracted. The definition must answer the question: what is the concept of information that fixes the phenomenon of information, sets it apart from energy and establishes distinction among forms of matter, energy and messages. It has been explained that the problem of the concept and phenomenon of information may be looked upon as the *general identification problem* of all elements. The problem of identification among objects in terms of varieties arises under both the principles of uniformity and ontological commonness. This means that there is an information problem in terms of content and phenomenon at all levels of existence. The conditions of two factors are identified for abstracting the General Information Definition (GID). The first one is the *characteristics* of elements in the universal object set that provide the *characteristic disposition* under relational continuum and unity in the qualitative-quantitative duality within the dynamics of actual-potential polarity. The second one is the *attribute signal* elements in the universal characteristic set that provide conditions of awareness in terms of *signal disposition* under relational continuum and unity in the qualitative-quantitative duality within the dynamics of actual-potential polarity. The dynamics of actual-potential polarity concerns the creation and destruction of varieties and categories as has been discussed in [R17.15, 17.16, R4.10], where analytical foundations of socio-natural transformations are presented with a search for qualitative mathematics under the general principle of opposites composed of socio-natural polarities and dualities with relational continuum and unity. The characteristic disposition and the signal disposition provide necessary and sufficient conditions to construct the *General Information Definition* (GID).

Definition 3.2.3.1 General Definition of Information (GID)—Symbolic
The General Information Definition (GID) $\mathbb{Z}$ is a Cartesian product of the form $\mathbb{X} \otimes \mathbb{S}$ where $\mathbb{X} \neq \emptyset$ is the characteristic disposition (set) and $\mathbb{S} \neq \emptyset$ is the attribute signal disposition (set) given the universal object set $\Omega \neq \emptyset$ and may be written as:

$$\mathbb{Z} = \{(x,s)|x \in \mathbb{X}, s \in \mathbb{S}, \forall \omega \in \Omega, \} = \{\mathbb{X} \otimes \mathbb{S}|\Omega \text{ and } (\mathbb{X} \otimes \mathbb{S}) \neq \emptyset\}$$

The General Information, therefore is a non-empty set of *characteristics* called characteristic disposition that provide a non-empty *signal set* called *signal disposition* which provides signaling evidence regarding the existence and identity of the elements in the universe in an objective sense from the ontological space. It is also a set of *relations* that creates awareness possibilities among ontological objects in a subjective sense in the epistemological space for all those ontological objects that have capacity to operate in the epistemological space. At the level of the ontological space every $x \in \mathbb{X}$ is objective and well-formed and every $s \in \mathbb{S}$ is also objective and well-formed. At the level ontology, the concepts and the corresponding problems of meaningfulness and truthfulness of the elements in $\mathbb{S}$ and $\mathbb{X}$ do not arise since the ontological space is the *organic identity* as well as the collection of all identities that fix varieties. For every characteristic $x \in \mathbb{X}$ there is one and only one corresponding signal $s \in \mathbb{S}$. The universe is unconceivable if $\mathbb{X} = \emptyset = \mathbb{S}$. The infinitude of the universe is such that $\#\Omega = \infty$ and $\#\mathbb{X} = \#\mathbb{S} = \infty^\infty$ where the symbol ∞^∞ is infinity of infinity.

3.3 The Set-Theoretic Structure, Gid, Variety and Category

The General Information Definition (GID) has been provided and its essential properties stated. The definition dwells on the relational structure of matter and energy through characteristics as properties of matter and signals as an energy process, both of which are related to conditions of distinction, commonness, variety and non-variety that present the identification problem of *categorial difference* and *categorial similarity*. The identification problem is present in all areas of knowledge production, cognitive operations and decision-choice activities. It was argued that the infinite universal object set Ω contains different ontological objects. These objects are seen in terms of varieties and groups of varieties called categories. It is useful now to present the set-theoretic structure of the ontological varieties and categories. The process of object-formation and category-formation is on the basis of analytical partitioning through characteristic analytics of differences similarities and clustering of the universal object set, where this object set is a set of universal varieties and categorial varieties. In other words, it uses a methodological decomposition and composition in characteristic-signal analytics

Definition 3.3.1: Partitioned Characteristic Set of the Universal Characteristic Set
The partition of the universal characteristics set $\mathbb{X}$ is the collection of non-empty groups of attributes that give sameness and difference, where such sameness and differences impose groupings or categories on the elements in the universal object space Ω. A partitioned characteristic set, $\mathbb{X}_\ell$ is a collection of attributes, $x_{\ell j}$ about any fixed $\omega_\ell \in \Omega$ such that $\mathbb{X}_\ell = \left\{ (x_1, x_2 \cdots x_j \cdots) \,|\, j \in \mathbb{J}_\ell \subset \mathbb{J}^\infty \right.$ and ℓ is *fixedin* $\mathbb{L}^\infty \}$, where $\mathbb{J}_\ell$ is a finite index set of attributes that defines the identity of variety $\omega_\ell \in \Omega$, with $\mathbb{X} = \bigcup_{\ell \in \mathbb{J}^\infty} \mathbb{X}_\ell$ and $\mathbb{J}^\infty = \bigcup_{\ell \in \mathbb{L}^\infty} \mathbb{J}_\ell$.

The path and the structure of the partitioned characteristic set in presenting the rise of varieties and necessary conditions of the characteristic-signal disposition for the abstraction of a General Information Definition are illustrated as a cognitive geometry in Fig. 3.5. It shows the nature of objective existence in terms of General Information Definition of varieties, distinction, similarities and differences. The definitional process of Fig. 3.5 is such that first we have a set-to-subset mapping followed by a set-to-object mapping, where the concept of a real definition relative

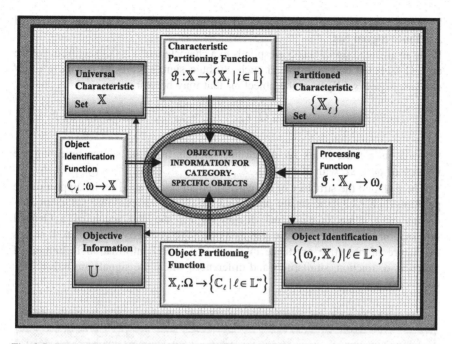

Fig. 3.5 The cognitive path for defining objective information, where $\mathbf{P}_1$ induces a partition on the universal characteristic set into sub-sets providing necessary conditions for the objective identification by **J**-function where the partitioned characteristic set is the used to induce a partition on the universal attribute signal set. These are the necessary conditions for the General Information Definition (GID) called the objective information square $\mathbb{U}$

to other forms of definition finds a scientific residence and importance [R3.19, R3.33, R3.40, R3.75, R4.13]. The analytical structure sees the elements of the universal object set, the *ITS* as objective. This is followed by a set-to-subset mapping which is completed by a point-to-set mapping as an illustrative path of the definition of *objective information*. The definitional process and the mappings are in the quality-time space for inter-categorial existence which allows the study of qualitative transformations in terms of qualitative motion rather than quantity-time space for intra-categorial existence that allows the study of quantitative motion, given the qualities of elements. The definitional process is constructed on the existence of the idea of quality-quantity duality under the principle of relational continuum and unity, where behind every qualitative disposition, there is a quantitative disposition and vice versa to form a unity of information.

The conditions of the *partitioned characteristic set* are necessary but not sufficient for informing, learning and knowing of the varieties and categories by other elements in the universal set. The sufficient conditions require an associated partitioned attribute signal to set which we now turn our attention.

Definition 3.3.2: The Partitioned Attribute Signal Set of the Universal Attribute Signal Set

If s is an attribute signal that corresponds to a characteristic $x \in \mathbb{X}_\ell$ sent from source $\omega_\ell \in \Omega$ then the collection of all such attribute signals from $\mathbb{X}_\ell$ is a *partitioned attribute signal set* $\mathbb{S}_\ell$ from the object ω_ℓ that may be written as:

$$\mathbb{S}_\ell = \left\{ \left(s_{\ell 1}, s_{\ell 2}, \cdots, s_{\ell j}, \cdots \right) \middle| \omega_\ell \in \Omega, \mathbb{X}_\ell \subset \mathbb{X}, j \in \mathbb{J}_\ell \subset \mathbb{J}^\infty \text{ and } \ell \text{ is fixed in } \mathbb{L}^\infty \right\},$$

where $\mathbb{J}_\ell$ is a finite index set of attribute signals from the source revealing the identity of $\omega_\ell \in \Omega$, with $\mathbb{X} = \bigcup_{\ell \in \mathbb{J}^\infty} \mathbb{X}_\ell$, $\mathbb{S} = \bigcup_{\ell \in \mathbb{J}^\infty} \mathbb{S}_\ell$ and $\mathbb{J}^\infty = \bigcup_{\ell \in \mathbb{L}^\infty} \mathbb{J}_\ell$.

The path and the structure of the partitioned attribute signal set in support of the partitioned characteristic set for presenting the rise of varieties and sufficient conditions of a characteristic-signal disposition for the abstraction of a General Information Definition are illustrated as a cognitive geometry in Fig. 3.6. It shows the nature of objective existence in terms of General Information Definition of varieties, distinction, similarities and differences where the signals are induced by the energy process.

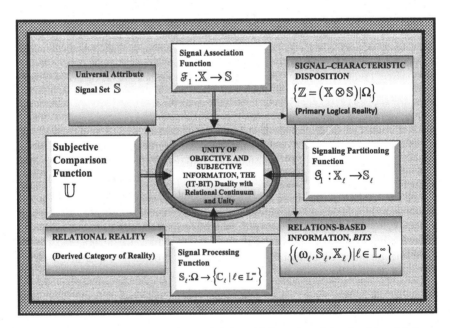

Fig. 3.6 Relational unity between characteristic-based information of the signal-based information in distinguishing varieties of ontological objects and information relation from epistemic reality in a relation to the universal object set

Note

The partitioned sets $\mathbb{S}_\ell$ and $\mathbb{X}_\ell$ are maximal finite sets with $\#\mathbb{S}_\ell = \#\mathbb{X}_\ell$ as well as $(\mathbb{S}_\ell \otimes \mathbb{X}_\ell) \subset (\mathbb{S} \otimes \mathbb{X})$ where each set-pair $(\mathbb{S}_\ell \otimes \mathbb{X}_\ell) = \mathbb{Z}_\ell$ provides a General Information Definition (GID) for informing, learning and knowing about $\omega_\ell \in \Omega, \forall \ell \in \mathbb{J}^\infty$. The sets $\mathbb{S}_\ell$ and $\mathbb{X}_\ell$ and the Cartesian product set $(\mathbb{S}_\ell \otimes \mathbb{X}_\ell) \subset (\mathbb{S} \otimes \mathbb{X})$ are well-formed (wf.), complete and exact such that each $\mathbb{Z}_\ell = (\mathbb{S}_\ell \otimes \mathbb{X}_\ell) \subset (\mathbb{S} \otimes \mathbb{X})$ is a perfect information in their natural setting. From the definitions of partitioned characteristic set and the partitioned attribute signal set, the nature of the *objective universe*, $\mathbb{U}$ may be defined as a schedule in terms of the *universal object set*, Ω the *universal characteristic set* $\mathbb{X}$, and the universal attribute signal set $\mathbb{S}$ to meet the conditions of categorial partitioning, where *ontological reality* appears in the categories of ontological objects as the *ITS*. It is useful to keep in mind that the objective universe is conceptualized by infinite objects, infinite characteristics and infinite attribute signals, where $\mathbb{U} = (\Omega \otimes \mathbb{X} \otimes \mathbb{S})$ and $\mathbb{Z} = (\mathbb{X} \otimes \mathbb{S})$ is the general information about Ω, and $\mathbb{Z}_\ell = (\mathbb{X}_\ell \otimes \mathbb{S}_\ell)$ is the general information about $\omega_\ell \in \Omega$. Required conditions about ontological varieties $\omega \in \Omega$ are now available to define categories and categorial realities in the universal object set. The conditions are such that $\omega_\ell \in \Omega$ is an ontological variety and $\mathbb{Z}_\ell = (\mathbb{X}_\ell \otimes \mathbb{S}_\ell) \subset \mathbb{Z}$ is its information, where $\mathbb{X}_\ell$ is the content-set and $\mathbb{S}_\ell$ is the signal-set. The idea of well-formed and objective information about a variety is claimed only for ontological elements that constitute the identities. This cannot be claimed without qualification for epistemological varieties.

Definition 3.3.3: Category of Realities and Categorial Variety

The categories of reality or simply an ontological category $\mathbb{C}_\ell$'s are collections of all identical elements (varieties) $\omega_\ell \in \Omega$ where each of the ℓth category is identified by a partitioned characteristic set, $\mathbb{X}_\ell$ in the form:

$$\mathbb{C}_\ell = \left\{ (\omega_\ell, x_{\ell j}, s_{\ell j}) \mid j \in \mathbb{J}_\ell \subset \mathbb{J}^\infty, \omega_\ell \in \Omega, \, x_{\ell j} \in \mathbb{X}_\ell, s_{\ell j} \in \mathbb{S} \text{ and } \ell \text{ is fixed in } \mathbb{L}^\infty \right\}$$

where $\mathbb{X}_\ell$ is the set of full attribute conditions called the characteristic disposition and $\mathbb{S}_\ell$ is a set of full signal conditions called signal disposition of a particular reality, $\omega_\ell \in \Omega$ Similarly, it may be written as $\mathbb{C}_\ell = \{(\omega_\ell, \mathbb{X}_\ell, \mathbb{S}_\ell) \mid \ell \in \mathbb{L}^\infty\}$, where $\mathbb{X}_\ell = \{(x_1, x_2 \cdots x_j \cdots) \mid j \in \mathbb{J}_\ell \subset \mathbb{J}^\infty, \ell \in \mathbb{L}^\infty\}$ and $\mathbb{S}_\ell = \{(s_1, s_2 \cdots s_j \cdots) \mid j \in \mathbb{J}_\ell \subset \mathbb{J}^\infty, \ell \in \mathbb{L}^\infty\}$ with $\mathbb{Z}_\ell = (\mathbb{X}_\ell \otimes \mathbb{S}_\ell) = \{(x_j, s_j) \mid j \in \mathbb{J}_\ell \subset \mathbb{J}^\infty, \ell \in \mathbb{L}^\infty\}$ providing GID for all $\omega_\ell \in \Omega$.

Definition 3.3.4: Ideticallity Relation

Given the general objective set Ω, two elements $(\omega_i, \omega_j \in \Omega)$ with the corresponding $\mathbb{Z}_i, \mathbb{Z}_j \subset \mathbb{Z}$ are said to be of the same variety if they satisfy the *identicallity relation*, written as $(\approx)$, provided that their information subsets meet the condition $\mathbb{Z}_i \subseteq \mathbb{Z}_j \leftrightarrow \mathbb{Z}_j \subseteq \mathbb{Z}_i$.

Note

An identicallity relation is an equivalence relation that divides the universal object set into equivalent classes which are pairwise fuzzy disjoint, called fuzzy equivalence classes $\mathbb{C}_\ell$, with common information subset $\mathbb{Z}_\ell = (\mathbb{X}_\ell \otimes \mathbb{S}_\ell)$. The identicallity relation is reflexive, symmetric and transitive with the condition, $\mathbb{Z}_i \not\approx \mathbb{Z}_j \Rightarrow (\mathbb{X}_i \otimes \mathbb{S}_i) \not\approx (\mathbb{X}_j \otimes \mathbb{S}_j)$. The same varieties are said to be categorial indifference. Given the general objective set Ω, any two elements $(\omega_i, \omega_j \in \Omega), \forall i, j \in \mathbb{R}^+$ with the corresponding $\mathbb{Z}_i, \mathbb{Z}_j \subset \mathbb{Z}$ are said to be of different varieties if $\mathbb{Z}_i \not\approx \mathbb{Z}_j \Rightarrow (\mathbb{X}_i \otimes \mathbb{S}_i) \not\approx (\mathbb{X}_j \otimes \mathbb{S}_j)$ with $\omega_i = v_i \in \mathbb{V}_i$ and $\omega_j = v_j \in \mathbb{V}_j$ with $\mathbb{V}_j \not\approx \mathbb{V}_j$.

Postulate 3.3.1: Objective Universe

The objective universe, $\mathbb{U}$ is an exhaustive, mutually exclusive and infinite collection of categories, $\mathbb{C}_\ell$, whose elements appear as schedules in the form $\mathbb{U} = \{\mathbb{C}_\ell | \ell \in \mathbb{L}^\infty\} = \{(\omega_\ell, \mathbb{X}_\ell) | \ell \in \mathbb{L}^\infty\}$. It is the collection of all the ontological objects (*ITS*) in categories of qualitative varieties in the ontological space. The objective universe is the same thing as the universal object set Ω the elements of which appear as varieties.

Postulate 3.3.2: Category

Given the universal object set, Ω, characteristic set, $\mathbb{X}$, and signal set, $\mathbb{S}$, let $(\approx)$ be an *identicality* relation defined over $\mathbb{X}$ and $\mathbb{S}$ then the group, $\mathbb{C}_k$, is said to be a category if and only if there exist $(\omega_1, \omega_2, \ldots, \omega_k) \in \Omega$ such that $(\mathbb{X}_1 \approx \mathbb{X}_2 \approx \ldots \approx \ldots \mathbb{X}_k) \in \mathbb{X}$, and $(\mathbb{S}_1 \approx \mathbb{S}_2 \approx \ldots \approx \ldots \mathbb{S}_k) \in \mathbb{S}$ then $(\omega_1, \omega_2, \ldots, \omega_k) \in \mathbb{C}_k$. Alternatively, if there exist $(\omega_1, \omega_2, \ldots, \omega_k) \in \mathbb{C}_k$, then the corresponding characteristic sets (characteristic disposition) are identical in the sense that, $(\mathbb{X}_1 \approx \mathbb{X}_2 \approx \ldots \approx \ldots \mathbb{X}_\ell) \in \mathbb{X}$ and so also are the signal sets (signal disposition) in the form $(\mathbb{S}_1 \approx \mathbb{S}_2 \approx \ldots \approx \ldots \mathbb{S}_k) \in \mathbb{S}$ with identical information of the form $\mathbb{Z}_k = (\mathbb{X}_k \otimes \mathbb{S}_k) = \{(x_j, s_j) | j \in \mathbb{J}_k \subset \mathbb{J}^\infty, k \in \mathbb{L}^\infty\}$ that defines a characteristic-signal disposition in the quality-time space holding qualitative disposition constant, and a characteristic-signal disposition in the quantity-time space holding qualitative disposition constant.

Note

The composition of the categories may be constructed with categorial indicator function of the form, $\mathbb{I}_\mathbb{C}(\omega), \omega \in \mathbb{C} \subset \Omega$. The notion of fuzzy equivalence classes are fuzzy categories in the sense that the crisp categories are formed by the combination of fuzzy decomposition and fixed-level cut. The elements $\omega_i \in \Omega$ are also called varieties and $\mathbb{C}_k$ is the kth-category. The categories may also be formed with variety indicator function $\mathbb{I}_\mathbb{C}(\omega)$ where given Ω, the entry to every category $\mathbb{C}$ is specified by a membership characteristic function that presents a distribution of degrees of belonging of the form $\mu_\mathbb{C}(\cdot)$. In this respect, we obtain a fuzzy indicator function that generates categories of the form:

$$I_{\mathbb{C}}(\omega) = \begin{cases} 1, & \text{if } \mu_{\mathbb{C}}(\omega) \in (\alpha, 1] \Rightarrow \omega \in \mathbb{C} \subset \Omega \\ 0, & \text{if } \mu_{\mathbb{C}}(\omega) \in [0, \alpha) \Rightarrow \omega \in \mathbb{C} \subset \Omega \end{cases},$$

where α is a solution to an optimal degree of beloning.

Postulate 3.3.3: Universal Object Set
The universal object set is a partition with respect to $\mathbb{C}$-categories such that each category is non-empty collection of identical varieties, $\mathbb{C}_\ell \neq \emptyset$, where the general intersection of all categories is an empty set $\cap_{\ell \in \mathbb{L}^\infty} \mathbb{C}_\ell = \emptyset$, $\forall \ell$ and for *any* $i \neq j \in \mathbb{L}^\infty$, $\mathbb{C}_i \cap \mathbb{C}_j = \emptyset$, with $\cup_{\ell \in \mathbb{L}^\infty} \mathbb{C}_\ell = \Omega = \mathbb{U}$. Each of the categories formed with fuzzy conditionality fuzzy is a fuzzy category and the collection of them is a family of fuzzy categories such that the conditions $\cap_{\ell \in \mathbb{L}^\infty} \mathbb{C}_\ell \neq \emptyset$, $\forall \ell$ and for any $i \neq j \in \mathbb{L}^\infty$, $\mathbb{C}_i \cap \mathbb{C}_j \neq \emptyset$ establish overlapping categories with a relational continuum with further conditions of $\cup_{\ell \in \mathbb{L}^\infty} \mathbb{C}_\ell = \Omega = \mathbb{U}$ that establish relational unity.

It may be noticed that corresponding to each element, $\omega_\ell \in \Omega$ there is a set of attributes, $\mathbb{X}_\ell$, a set of signals $\mathbb{S}_\ell$ and information $\mathbb{Z}_\ell = (\mathbb{X}_\ell \otimes \mathbb{S}_\ell)$ that identifies it. The collection of all $\omega_\ell \in \Omega$ with attributes $\mathbb{X}_\ell$ and signals $\mathbb{S}_\ell$ constitutes a category $\mathbb{C}_\ell$ and the collection of all these categories constitutes the objective universe $\mathbb{U}$. The universal object set Ω is simply the objective universe $\mathbb{U}$ without the defining characteristic set $\mathbb{X}_\ell$, signal set $\mathbb{S}_\ell$ and information $\mathbb{Z}_\ell$ that partitions Ω into categories of entities. Thus $\Omega = \{\mathbb{C}_\ell | \ell \in \mathbb{L}^\infty\} = \{(\omega_\ell, \mathbb{X}_\ell, \mathbb{S}_\ell) | \ell \in \mathbb{L}^\infty\}$ where $\mathbb{X} = \cup_{\ell \in \mathbb{J}^\infty} \mathbb{X}_\ell$, $\mathbb{S} = \cup_{\ell \in \mathbb{J}^\infty} \mathbb{S}_\ell \mathbb{J}^\infty = \cup_{\ell \in \mathbb{L}^\infty} \mathbb{J}_\ell$ and $\#\mathbb{X}_\ell = \#\mathbb{J}_\ell = \#\mathbb{S}_\ell$. Both $\mathbb{U}$ and Ω represent the collection of the *primary categories of reality*. Definitions (3.3.1–3.3.3) specify the objective existence of entities, states, processes and events in their varieties that constitute the sources of a *characteristic-signal-based information set* which we have referred to as the objective information in terms of ontological characteristic-signal disposition of the form $\mathbb{Z}_\ell = (\mathbb{X}_\ell \otimes \mathbb{S}_\ell)$. The sources present three important items of the universal object set Ω, the universal characteristics set $\mathbb{X}$ and the universal signal set $\mathbb{S}$. The universal object set is the primary category of existence while the universal characteristic and attribute sets are derivatives by methodological constructionism and reductionism. The three sets are considered as *factual reality* by GID in the sense that their existence is independent of awareness of any object in the universal object set. Thus the characteristic-signal based information is the expression of the *objective reality* about the universal objects. Furthermore, the universal object set is infinitely closed under *category formation* at any given cosmic or cosmological time. It is also closed under *continual transformation* where *categorial conversion* presents the necessary conditions [R17.15] and *Philosophical Consciencism* under decision-choice systems presents the sufficient conditions for the directions and the nature of transformations [R17.16].

The components of the characteristic disposition in the GID deals with exactness and completeness of the credibility of the source. The component of the signal disposition in the GID deals with the subjective aspects of information as may be

received and interpreted by other elements in the universal object set. These subjective aspects of information are intimately related to the interpretation and
response to the characteristic disposition. We begin with the observation that every
element in the universal objects set sends attributive signals that correspond to the
set of attributes which defines its identity and hence its variety. The attribute signals
create conditions for awareness by ontological objects in terms of relations between
the source objects and recipient objects in the universal object set. Every element
(*IT*) in the universal object set is both a source object and a recipient object in the
sense that it sends and receives information through signals that create the capacity
to assign differences and similarities to the nature of the ontological varieties and
categories in the quantity-quality duality, and to form judgments about elements in
the dynamic behavior in the actual-potential polarity. In this way the
source-recipient modules establish relationships that are defined by informing,
learning and knowing and also decision-choice actions about ontological reality. It
is these conditions that help to establish the categorial indicator function and the
required category formation. The GID establishes *sufficient conditions* for the
solution to the general *identification problem* in the *ontological statics* as well as the
initial necessary conditions for the study of transformations of varieties and categorial varieties

3.4 The Summary of the General Information Definition (GID)

On the basis of the discussions and the definition of information, it is useful to state
the basic requirements of any information definition that the GID may engender
over the knowledge production space. From these requirements it will be shown
that any definition of information is a derivative of GID in the sense that it projects
distinction, difference and similarity through a distribution of varieties where each
variety is established by a characteristic-signal disposition.

Let $\mathbb{W}_\ell$ be an information structure from source $\omega_\ell \in \Omega$ to destination $\omega_j \in \Omega$,
then $\mathbb{W}_\ell$ is an instance of GID if:

1. $\mathbb{W}_\ell$ consists of a maximal characteristic set $\mathbb{X}_\ell$ where $(\#\mathbb{X}_\ell) \geq 1$ with a corresponding maximal signal set $\mathbb{S}_\ell$ and $(\#\mathbb{S}_\ell) = (\#\mathbb{X}) \geq 1$ with $\#(\mathbb{X}_\ell \otimes \mathbb{S}_\ell) \geq 1$ as
   the maximal characteristic-signal disposition.
2. The all characteristic-signal dispositions of the form $\mathbb{Z}_\ell = (\mathbb{X}_\ell \otimes \mathbb{S}_\ell)$ are
   well-formed.
3. Each well-formed characteristic-signal disposition, $\mathbb{Z}_\ell = (\mathbb{X}_\ell \otimes \mathbb{S}_\ell)$ from $\omega_\ell \in$
   Ω to $\omega_j \in \Omega$ is meaningful to all $\omega_j \in \Omega$ with cognitive capacity
4. The well-formed characteristic-signal dispositions of the form $\mathbb{Z}_\ell = (\mathbb{X}_\ell \otimes \mathbb{S}_\ell)$
   from the sources $\omega_\ell \in \Omega$ to the destination $\omega_j \in \Omega$ are objective, complete and
   exact in transmission such that there are no uncertainties, accidents and risks.

The characteristic-signal disposition is the identity and the primary category of information definition that establishes the phenomenon of information. All other information definition may be shown to be logical derivatives by methodological reductionism. The channel of transmission is *noiseless* around which it may be shown how information imperfections arise through the logical movement from the ontological space to the epistemological space. The general definition of information is what is also called *ontological information* in this monograph. The onto-logical information defines the ontological space by fixing the morphology of the elements in the universal object set. The study of the morphology of these elements is the study of the ontological information as captured by characteristic-signal dispositions that specify the varieties and categories of elements of *what there is* (the actual) and *what would be* (the potential) within the dynamics of actual-potential polarity, where some old varieties may be destroyed and new varieties created under the conditions of continual differentiation.

3.4.1   The Objective of the General Theory of Information: A Non-standard Approach

The *theory of general and non-standard information* seeks a general framework of understanding information, and information as input into knowledge production such that the results become inputs into decision-choice processes of management of command-control systems for continual creations of varieties and categories in socio-natural actual-potential polarities. The first task is to give a definition of information that is general (GID) which will serve as the primary category of definitional foundation from which derived definitions can be shown to emerge by methodological constructionism or reducible to it by methodological reductionism. This general information definition fixes the boundaries of the phenomenon of information as compare to the phenomena of matter and energy. It also establishes the idea that information is a property of matter and is also a property of energy by categorial derivative from matter. The GID must contain the concept and content of information in the quality-quantity duality with relational continuum and unity where information is coextensive with matter and energy. At the ontological space, the measure of information is the size of the characteristic set and the corresponding possibilistic set of transformation of varieties. It is this possibilistic set that trans-lates to a probabilistic set in the epistemological space. The relational structure between the possibilistic set and the probabilistic set in cognition and decision-choice actions must be clearly understood from the information structure. The possibilistic set establishes necessary conditions and projects necessity while the probabilistic set establishes sufficient conditions and projects freedom, where possibility is to necessity and probability is to freedom over the epistemological space. The ontological information is contained in the material conditions, broadly defined. The derived process of GID is shown in Fig. 3.7.

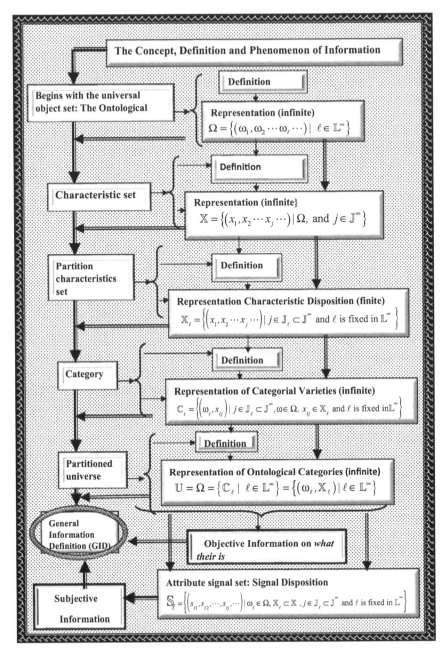

Fig. 3.7 Conceptual system of information definition and representation

3.4.2 Information, Knowledge and Decision-Choice Systems in the Ontological Space

Information is that thing which reveals distinction, commonness, diversity, differences and similarities of universal existence. This thing is revealed by the differential characteristic sets of the source existence which informs the destination existence through the corresponding signal sets. It will be shown that the structure of the theory of non-standard information gives rise to the theory of information transmission, the theory of information measurement and the semantic theory of information as sub-theories when the ontological information is mapped onto the epistemological information to extend the General Information Definition. The non-standard theory of information will then incorporate the classical information theory which will be coextensive. In this respect, the non-standard theory of information constitutes a general theory of information. At the level of explanatory science, the non-standard theory of information defines and explicates the concept of information and fixes the boundaries of the phenomenon of information. It explains the content, transmission and measurement of information. At the level of prescriptive science, the non-standard theory of information specifies the conditions of continual information creation through the dynamics of categorial variety and continual transformation of matter and energy, and shows how information becomes input into decision-choice actions for continual transformation within the organic actual-potential polarity.

As it has been explained, at the level of ontology, the GID is perfect in the sense of completeness and exactness for both the source existence and destination existence. The nature of the perfectness of the GID is that the conditions of randomness and probability of processes and outcomes do not arise. Every variety is under exact creation with perfect information under a natural command-control decision-choice system. Information therefore, is said to be perfect if the characteristic-signal disposition is complete and exact with noiseless transmission for both the source existence and destination existence. The exactness and completeness of the information are such that one cannot speak of uncertainty, risk and accidents due to incompleteness of quantity of information neither can one speak of uncertainty, risk and accidents due to inexactness of quality of information. Possibility and probability do not exist and hence one cannot discuss the concept, phenomenon and measurement of information in terms of possibility and probability. The only things that exist are actual, potential and actual-potential processes induced by continual socio-natural transformations. All the socio-natural transformations are not only self-induced from within but are matter-energy-information processes such that *existence* and *variety* are connected to *matter, power* and *work* (motion) which are connected to *energy, and knowledge* and *decision* are connected to *information* while knowledge and decision are connecter to transformation of socio-natural varieties and categorial varieties. It is in this sense that energy and information are said to be derivatives of matter which is considered as the primary category of existence. It is also here that the information-decision-interactive process acquires an analytical power in understanding quantitative and qualitative motions. The conditions of the socio-natural transformation present themselves in terms of the *necessary conditions* and *sufficient conditions*. The necessary conditions for socio-natural transformation are discussed and critically examined in the *theory of categorial conversion* [R17.15] while the sufficient conditions are also discussed and critically examined in the *theory of Philosophical Consciencism* [R17.16, R17.18] and related to general conditions of necessity and freedom

As far as the inputs into command and control of decision-choice systems are concerned in the ontological space, information is knowledge and knowledge is information. The commands and controls of socio-natural decision-choice processes shape and mold the ontological environment where the decision-choice objectives and the degrees of utilization of matter and energy as well as the effectiveness of their usages depend on the ontological environment under the principles contained in info-dynamics. One may recall that by the principles inherent in the info-dynamics, information is continually being created to update the existing information stock and hence the information is always in stock and flow disequilibria in never ending processes of the dynamics of categorial variety [R17.18]. It is the existence of categorial varieties in terms of characteristic-signal dispositions that gives meaning to the concept and phenomenon of information for informing, learning, knowing and teaching to create inputs for socio-natural decisions and choices for the categorial conversion of varieties.

Without varieties, in terms of differences and commonness as expressed by characteristic-signal dispositions, there will be no decision-choice problems. Decision-choice problems arise as a result of information-knowledge processes regarding matter and energy. The decision-choice processes provide the dynamics of change and are intimately connected to the information-knowledge processes which in turn are intimately connected to the dynamics of categorial variety and continual socio-natural transformations which are then connected to the matter-energy processes. Here, we are confronted with an important problem regarding quantity of information. Should the unit of measure of information content at the source be the same as the unit of measure of information transmitted, and should it also be the same as the unit of information at the destination? This question is unimportant at the level of ontology since one is dealing with perfect information at the source existence, transmission process and the destination existence where maximum information from the source is transmitted to the destination in a noiseless medium as shown in Fig. 3.2. All adjectives attached to information, such as social information, economic information, political information, scientific information, geomorphological information, biological information chemical information and many others are processes to identify varieties of characteristics-signal dispositions. This is also the case of the use of adjectives and adverbs in languages.

Chapter 4
From the Ontological General Information Definition (OGID) To the Epistemological General Information Definition (EGID)

The discussions in Chap. 3 centered on definitional conditions of the concept and phenomenon of information at the ontological space. The analytical justification of this approach rests on the fact that by understanding the phenomenon of information at the level of *what there is*, a rational pathway is opened for examining and analyzing the conditions of information at the level of epistemology for generalization and integration of different approaches to the definition and phenomenon of information. The definition of information as a characteristic-signal disposition at the level of ontology is the most general that forms the basic foundation as the primary category of information definitions and phenomena from which the definitions of the concepts of knowledge, uncertainty, ignorance, possibility, probability, risk, accident, anticipation, expectation, progress, change and others are derived. In other words, the definition of all these concepts are traceable to the General Information Definition (GID) by the process of methodological reductionism. Every information at any level is made up by characteristic and signal dispositions that accommodate quality and quantity to establish the understanding of differences and commonness in varieties and categories that form the foundations of mathematical concepts of set, group and subgroups in time and over time. The whole notion of classification, cluster analysis and grouping such as tribe and ethnicity are linguistically meaningless without the existence of varieties and a general information definition that relates to the varieties.

All definitions of the concept and phenomenon of information at the level of epistemology are either derived from or reducible to the primary definition (GID) at the ontological level. This approach to the concept and phenomenon of information is consistent with the principles of statics and dynamics of categorial variety at both the levels of qualitative and quantitative dispositions. It is also consistent with the general principle of opposites composed of dualism (with excluded middle, generating binary and discreteness without contradiction in thought) and duality (with relational continuum and unity, generating analogue and continuity with acceptance of contradictions in thought).

K.K. Dompere, *The Theory of Info-Statics: Conceptual Foundations of Information and Knowledge*, Studies in Systems, Decision and Control 112, DOI 10.1007/978-3-319-61639-1_4

4.1   The Concepts of Epistemological Variety and Category

Most of the past and current approaches to the development of the theory of information have been from the epistemological space. Even over the epistemological space, there is a restriction where the source and destination are seen in terms of cognitive agents with some level of information processing capacity under the principle of communication. These approaches to the definition of the concept and phenomenon of information have created a number of limitations on the generality of the phenomenon of information and its function as input into the awareness-knowing process under the principle of the acquaintance-learning process. There is a whole confusion about the discussions on the concept and phenomenon of information as seen from the epistemological space. The confusion over the phenomenon of information, its concept and content has led some scientific thinkers to claim that information is not matter or energy. *Information is information, not matter or energy. No materialism which does not admit this can survive at the present day* [R20.18, p. 155]. To accept or reject the claim that information is not matter or energy, one must have a defining concept of the phenomenon of information and then relate it to defining concepts of matter and energy. The position in this monograph is that the universe is composed of organic categories of matter, energy and information where information is a property of matter. It is also a property of energy by categorial derivative as well as a property of information by categorial derivative. Matter, energy and information exist as organic varieties in relational continuum and unity. Without matter and energy, there is no information and without information matter and energy are indistinguishable and recognizable.

At the level of knowledge-production over the epistemological space, one must answer a series of questions. Is energy matter or is matter energy? Alternatively, is the existence of the energy a derivative of the existence of matter or vice versa? Is the existence of information derived from either the existence of matter or energy? Similarly, is the existence of matter or energy derived from the existence of information? Do matter, energy and information have an independent existence or do they have mutually dependent existence as well as are they mutually self-creating? These questions must be answered by those who claim that information is not matter or energy. To claim that information is not mater or energy is to claim their independent existence which will violate the principle of universal existence in relational continuum and unity. Can information exist by itself without the existence of either matter or energy? In other words, can information exist in vacuum and how will the vacuum be? At the level of cognition, these questions involve problems that are not different from the *mind-matter problem* in the philosophy of existence, where mind is unperceived and matter is perceived. Similarly, energy and information are unperceived and matter is perceived.

Analytically, the claim of energy as a derivative of matter initializes matter as the primary category of existence. Similarly, if one claims matter to be a derivative of energy one then initializes energy as the primary category of existence. How does one know the differences and similarities between energy and matter? Similarly, how does one know that information is information, and information is neither matter nor energy at the epistemological space of informing-knowing-learning activities to create inputs into command-control management of decision-choice system in either static or dynamic domain? Is the knowing of differences, similarities and commonness not through varieties that are established as alternatives by categorial characteristics as reflected through the distribution of signal dispositions as the distribution of characteristic dispositions? Is the knowledge of alternatives not established by characteristics and informed by signals? It is always possible at the level of epistemology to see the concept and definition of information in diverse forms. Each form of seen information at the level of epistemology is an illumination on a particular human problem whose structure and solution are being examined in a particular branch of the human thought system about the universal existence for cognition and decision. The categorial variety presents itself as qualitative varieties that are expressed as *inter-categorial differences* and also as quantitative varieties that are expressed as *intra-categorial differences* for comparative analysis from without and from within. One thing that is analytically and practically clear is that the utility of information finds expression in defining the capacity for establishing recognizable differences, similarities and commonness for the distribution of the varieties of all kinds of messages over the epistemological space and to facilitate decision-choice activities over the space of alternative varieties.

The epistemic position that is being taken in the development of non-standard and general information theory in this monograph is that any form of seeing the phenomenon of information at the level of epistemology is reducible to conditions of the ontological state in which matter and energy reside. It is here that information assumes the properties of matter where the finitude or infinitude of information is related to the specificity and universality of matter. The finitude relates to variety while the infinitude relates to the family of families of infinite varieties. It has been argued in terms of how information is derived from matter and as a property of matter in the ontological space through the establishment of the GID (General Information Definition). To say that information is information is simply a circularity of thought that says nothing substantive and brings no illumination to the system of informing and knowing. To say information is not matter or energy, without saying what it is provides us no conditions of understanding and judgment. Information can exist unperceived and uncommunicated at the level of epistemology. The lack of the perception of information at the epistemological space does not deny its ontological existence. The relational structure of matter, energy and information is complex because of their continuum and unity in the universal existence. For just as matter is coextensive with the universe, so also is information coextensive with matter. Just as energy can be shown to emerge out of matter by a process so also can information be shown to emerge from matter through the interplay of the duals in the quality-quantity process by categorial conversion. Information has no independent existence from matter and energy.

The logical dynamics of categorial conversion of matter recognizes categorial varieties and categorial differences at the level of ontology and are translated to alternatives at the level of decision-choice actions. It accepts matter, energy and information as varieties of the same ontological existence by placing distinction as well as presenting an explanation as to the rise of differences and distribution as parts of the processes, but not as irreducible fundamentals in a separate existence. Just as energy is reducible to matter by a process so also is information reducible to matter by a process. At the level of epistemology, the process of thought keeps in mind the universal principle of opposites where there is the quality-quantity duality, in terms of characteristics, with relational continuum and unity of all varieties as already been discussed in [R44.13, R17.15, R17.16]. Logically, the ontological universe is seen as a collection of different processes of the same existence with temporary equilibrium states that allow recognitions of varieties of states and processes. Here, a clear logical observation must be made. To say that energy and information are derivatives of matter does not allow one to conclude that information and energy have mass that is attributed to matter. It simply says that the universe of matter is closed under categorial conversion with a continually evolving number of varieties and that nothing can be informed or known without an anchorage on matter. The varieties are the internal works of matter with a relational complexity of information and energy. The collection of categorial varieties in the ontological space generates the *ontological information*. How is this ontological information seen and how does it help to understand the concept and phenomenon of information at the level of epistemology?

4.1.1 Epistemological Variety, Matter, Energy and the General Information Definition (GID)

As one moves from the concept and phenomenon of information at the ontological level to the concept and phenomenon of information at the epistemological level, one is confronted with an interpretive problem of the elements in the signal disposition and the formation of conditions of epistemological reality in comparison to the ontological reality. The information at the ontological space is called the *ontological information* that is formed by *ontological characteristic-signal disposition* that allows the understanding of ontological varieties. The information at the epistemological space is called the *epistemological information* that is formed by *epistemological characteristic-signal disposition*. Are the concept and phenomenon of information at the ontological level the same as the concept and phenomenon of information at the epistemological level? If the answer is affirmative, then how does one demonstrate them as one moves from ontological space to the epistemological space? This will require that one shows that the epistemological characteristic-signal disposition is a derivative from the ontological characteristic-signal disposition in all informing-knowing-learning processes. If the answer is negative, then what are the differences and how does one develop the concept and phenomenon of information at the epistemological space in which the difference may be demonstrated. An auxiliary problem is to demonstrate how this epistemological information developed from different foundations allows one to be informed as to *what there is*, the ontological objects at any point in time as *info-statics*, and *what would be* over time as *info-dynamics*. In this monograph, the development of epistemological information proceeds from the developmental foundations of the ontological information that have been developed and discussed in Chap. 3.

There is a premise from the conditions of ontological space that matter, energy and information are interconnected and are inter-reducible categories of universal existence. The interconnectedness and inter-reducibility of these organic categories and continual creation and destruction of varieties produce relational continuum and unity of universal existence the knowledge of which can only be revealed by the interplay of characteristics and signals that form the informing process. In other words, matter, energy and information are not only connected but form relational continuum and unity of categorial existence. The conceptual tools for the analysis of the inter-reducibility and categorial unity in the epistemological space are those contained in nominalism, constructionism and reductionism [R4.13]. The concepts of nominalism, constructionism and reductionism must be related to the task of establishing categorial unity and inter-reducibility at the level of epistemology. In building the relationality of the concepts of ontological information and the concepts of epistemological information, it is useful to understand that in *methodological nominalism*, one holds the characteristics of concrete existence as real, and then as the primary category of existence from which other categories are abstracted as derivatives or to which other categories of existence are reduced.

In terms of the current *matter-information problem* of inter-connectedness and inter-reducibility process, one holds the characteristics of matter as the primary reality of existence and in terms of characteristic disposition from which signal disposition is a derivative. It must also be noted that the concept of signal disposition connects the ontological space to the epistemological space through acquaintance language formation and development in terms of informing, naming, knowing and learning. It is through this connection that cognitive agents develop knowledge about elements of ontological existence. By receiving the signal disposition, one develops a process in the epistemological space to identify the corresponding characteristic disposition and the corresponding object or variety in the ontological space. The process is a movement from ontological to epistemological through the signal disposition. It denies any movement from epistemological to ontological where epistemological characteristic disposition is created in the mind and one looks for the ontological signal disposition to justify ontological existence in the sense that what exists in the cognition may turn out not to be real but illusion. *Constructionism* is a forward logical process in the epistemic space to show how derived categories of varieties are pulled out from the primary category of existence. In terms of the current matter-information problem, one is to show how information is derived from the characteristics of matter as the primary category of existence. Reductionism is a backward logical process in the epistemological space to show how any derived category can be traced to the primary category of existence.

In terms of the current matter-information problem and the task at issue, one is to show how the concepts relating to the phenomenon of information are traced to the concepts that are proper to the concepts of characteristics of matter as the primary category of existence. It is here, that the general information definition (GID) as a characteristic-signal disposition establishes the power of solution to the identification problem of all varieties where messages are not independent of the varieties. As it has been argued, a message is carried by the signal disposition and the content of the message is the characteristic disposition, such that the information of a message is the characteristic-signal disposition. Every message is a variety that carries a specific information content different from other messages without which informing and knowing are impossible.

The relational structure of nominalism, constructionism and reductionism at the level of epistemology is equivalent to that which connects matter, energy and information in continuum and unity at the level of ontology. The driving force is the continual transformation of categories under the relational structure of primary and derived categories within the conversion actions in the actual-potential polarity operating with the analytical structure of a set of dualities and the principle of opposites. The epistemological space is cognitive creation to allow cognitive agents to connect to the ontological space for information activities in such a way that the epistemological space is a categorial derivative for knowing through acquaintance from the ontological space which constitutes the primary category. From the viewpoint of the *general definition of information*, the epistemological space is empty without the ontological space. The size of the epistemological space at any time is limited by cognition and is also constrained by limitativeness and limitationality of collective cognitive capacity. There are some important observations. In the ontological space, matter is considered as the organic primary category of existence and energy and information are considered as organic derived categories of existence for knowing, learning and teaching. In the epistemological space, methodological nominalism is considered as the organic primary category of logical existence and constructionism and reductionism are considered as derived logical categories of existence. The processes of informing, knowing and learning are such that the ontological information, carried by the *ontological signal disposition*, is linked to the development of epistemological information by decoding the ontological signal disposition through the methodological nominalism that constitutes the primary logical existence in the epistemological space. In this respect, nominalism is linked to matter and hence connects the definition of information at the epistemological space to the proposition of matter in the ontological space, where the methodological nominalism is supported by methodological constructionism and methodological reductionism for justification and continual acceptance of knowing.

The connection between ontology and epistemology is such that nominalism may be related to the development of information through language and communication in all aspects of ordinary and formal language structures at the level of epistemology on the basis of which the principles of methodological constructionism and reductionism are derived. The role of language may be seen as representation of signal dispositions that must be cognitively processed to create an epistemological characteristic disposition. Constructionism may related to epistemological development of thought systems that relate to the structure of the epistemological characteristic disposition with further development of abstract ideas that permit the growth of illusions, epistemic fantasies, logical contradictions and the knowing of reality within the information given languages. Reductionism, on the other hand, may be related to the examination of the validity of the constructed thought system as to how well it approximates ontological reality. In other words, methodological reductionism seeks to verify the relational equality between the epistemological characteristic disposition and the ontological characteristic disposition and then strike a difference called the *epistemological distance* [R4.10].

It is through the principle of acquaintance and the methodological nominalism that *epistemological information* must be defined and shown as a derivative of ontological information. The epistemological information then serves as input into the methodological processes of constructionism and reductionism for any given language to the extent to which the language is a representation of information. In this respect, nominalism, constructionism and reductionism connect the epistemological space to the ontological space for informing, knowing and learning to generate knowledge production as an input into the management of socio-natural command–control processes in the decision-choice systems for transformation of varieties and categorial varieties. Nominalism constitutes the primary logical category of existence for designing rules of naming, vocabulary, composition and meaning associated in the General Information Definition (GID) in the epistemological space, where the naming, vocabulary composition and meaning are derived from propositional conditions of ontological signal disposition to create the epistemological characteristic disposition and the associated epistemological signal disposition. The development of methodological nominalism is only possible under some principle of acquaintance.

In this analytical system of reasoning, there are epistemic differences and similarities among matter, energy and information as viewed in the epistemological space relative to the ontological space. These differences and similarities are categorial in nature that belong to logical grammar which exist in the epistemic process over the epistemological space. They are categorial differences and categorial similarities that are epistemological but not ontological in nature, in the sense that information and energy are completely the result of a process of critical internal organization and reorganization of matter to produce *varieties* and *categorial varieties*. The process of critical organization and reorganization are part of the principle of continual negation-of-negation transformation in the actual-potential polarity under the active hands of family of dualities with relational continuum and unity. The conceptual propositions about information just as those about energy are completely reducible to conceptual propositions about matter through the family of characteristic-signal dispositions within the infinite collection of quality-quantity dualities under the dynamics of an infinite collection of actual potential polarities.

Energy and information are derivatives of matter and hence retain the categorial differences and similarities where the basic universe is matter that constitutes its primary category of universal existence of matter, therefore, are the basic inputs into the continual transformation while the information is an input into the socio-natural processes of knowledge production whose results become inputs into the socio-natural management of command-control elements of decision-choice systems for the transformation dynamics of the family of socio-natural actual-potential polarities in the universal system. At the level of thought, there are differences between concepts and objects, between concepts and processes and between concepts and concrete existence that present categorial differences and similarities at the epistemological space where the concept of information and the object are embedded in the categorial difference within the object concept duality. The categorial differences arise from the encoding and decoding of the elements in the signal disposition as associated with the characteristic dispositions which reveals the multiplicity of the reality of existence.

4.1.2 Acquaintance, Epistemological Variety, and the General Information Definition (GID)

At the level of epistemology and with regard to the epistemological characteristics, varieties and categories are formed on the basis of awareness through acquaintances with ontological signal dispositions that present the capacity to form distinction, similarity, difference and commonness about ontological objects and categories through the corresponding characteristic dispositions which together constitute the ontological information as presented by the GID. It is on the basis of the ontological information that ontological objects, whether cognitive or non-cognitive agents, are informed about the existence of other ontological objects, and those with the capacity of cognition develop knowledge about their environment and learn from the ontological experience of the possibility to effect changes in their environment and take probabilistic action to make it happen. In this respect, the classical theory of information composed of the theory of information transmission, semantic theory of information, mathematical theory of communication, theory of coding, synthetic theory of information and statistical theory of information (quantity of information associated with signal disposition), is an attempt to understand the nature of information at the level of epistemology, where information as an energy process is a *relation* among ontological objects. Most of these approaches take information as implicitly defined by information or *data*.

Here, one is confronted with a scientific-philosophical problem of quality-quantity opposites with neutrality of time, where one may either adopt the *classical paradigm of methodological dualism* with excluded middle, relational separation and disunity, or one may adopt the *fuzzy paradigm of methodological duality* with relational continuum and unity for information representation, processing, analysis and computing [R4.7, R4.8, R4.9, R4.13]. Here, methodology is viewed in the scientific-philosophical domain as a cognitive framework in which linguistic computations are developed to create meanings from language in the process of knowledge development, learning, communication and teaching. This linguistic computation includes logic, methods and techniques of reasoning, analysis, synthesis and acceptance conditions of epistemic claims where language assumes the character of epistemological signal disposition constructed from the ontological signal disposition. The set of conditions for the use of the fuzzy paradigm rests on the recognition that the epistemological signal dispositions may deviate from the ontological signal dispositions through the process of acquaintance that may generate noisy channels of transmission in the sense of vagueness. The set of conditions for the use of the classical paradigm of thought also rests on the condition that the epistemological signal dispositions are the same as the ontological signal dispositions through the process of acquaintance that may generate small if any noisy channels of transmission in the sense of exactness.

4.1.3 Concept, Content and the Phenomenon of Information at the Level of Epistemology

The problem of information and the solution process become more complex as an attempt is made to examine how the General Information Definition developed in the ontological space may be extended to the epistemological space, and to examine other possible definitions and conditions of information in the epistemological space. The epistemological space is the space for the activities of cognitive agents whose actions are involved in information-decision-interactive processes under the conditions of intentionality. This intentionality may be formed under the principle of cost-benefit rationality defined within the cost-benefit duality with relational continuum and unity. The relational continuum and unity must be seen in terms of t a system where the costs and benefits are mutually create themselves in such a way that for every cost there is a benefit support and vice versa in relation to the resolution structure of the *Asantrofi-Anoma* problem [R17.15, R17.16, R17.19]. The extension of the General Information Definition to the epistemological space implies that from the ontological characteristic-signal disposition one must derive an epistemological characteristic-signal disposition that can serve as an extension of the GID in the epistemological space.

In relating the ontological characteristic-signal disposition, the actions of acquaintance on the part of cognitive agents are introduced while the behavior of non-cognitive agents are neglected. Here, the source is regarded as ontological but not epistemological. The content of the message is the ontological characteristics that are presented as varieties and categorial varieties of ontological objects and carried by signal dispositions to the epistemological space to create conditions of awareness through acquaintances. The awareness, therefore, is through acquaintances with the elements in the signal disposition which allows the elements in the characteristic disposition to be identified and related to a categorial variety. The source and the message from the ontological space are perfect. Any imperfection is attributed to the activities in the epistemological space that serves as the destination of informing, knowing, learning and teaching, where vagueness, inexactness and approximations are generated by cognitive limitation which are unavoidable in observation and representation. It is useful to note that the theory of acquaintance is about the problems of ontological varieties and categorial varieties, and mathematical constructs just as languages are about representation and extension of the results of acquaintances.

The imperfection in the destination creates noise which is epistemological but not ontological. The analysis of the epistemological noise must be viewed in terms of acquaintance with elements of the ontological signal disposition within the exactness-inexactness dualism and completeness-incompleteness dualism with excluded middle and relational disunity or within exactness-inexactness duality and completeness-incompleteness duality with relational continuum and unity. The analytical framework is such that the acquaintance process in the epistemological space may be segmented into cohorts in terms of exactness, inexactness, completeness and incompleteness of acquaintances with the ontological signal disposition. Here, one may specify the combination of exact and complete acquaintance, inexact and complete acquaintance, exact and incomplete acquaintance and inexact and incomplete acquaintance. The cohorts are shown in Table 4.1 in relation to awareness of the elements of the ontological signal disposition. The works in [R3, R3.6, R3.13, R3.14, R3.26, R18, R18.4, R18.42, R18.58, R18.63] and others may be useful for these discussions in relation to the nature of the degrees of vagueness and exactness in representation of the signal dispositions and their distributions over the epistemological space and prerequisite of choice of paradigm of thought.

Table 4.1 Exact-inexact duality with complete-incomplete duality

	EXACT ACQUAINTAN	INEXACT ACQUAINTANCE
COMPLETE ACQUAINTAN CE	EXACT AND COMPLETE ACQUAINTANC	INEXACT AND COMPLETE ACQUAINTANCE
INCOMPLETE ACQUAINTANCE	EXACT AND INCOMPLETE ACQUAINTANCE	INEXACT AND INCOMPLETE ACQUAINTANCE

4.1.3.1 Exact and Complete Epistemological Information from the Ontological Space

Let S be the set of the number of signals encountered in the process of acquaintance and $S \subseteq \mathbb{S} \subseteq S$ then $(\#S) = (\#\mathbb{S})$, and $[(\#S) - (\#\mathbb{S})] = 0$. In this respect, there is no observational deficiency in terms of quantitative categorial difference under conditions of exactness in the acquaintance of the elements in the signal disposition which, implies that the epistemological set X as formed is equal to the ontological characteristic set and hence $X \subseteq \mathbb{X} \subseteq X$ and $(\#X) = (\#\mathbb{X})$ with $[(\#X) - (\#\mathbb{X})] = 0$. The pair $(X \otimes S)$ representing the epistemological characteristic-signal disposition is the *epistemological information* and is equal to the ontological characteristic-signal disposition $(\mathbb{X} \otimes \mathbb{S})$ where there is no inexactness or vagueness and there is perfect acquaintance. The absence of inexactness translates to the conditions that the space of epistemic actions is perfect where the epistemological characteristic-signal disposition is exactly equal to the ontological characteristic-signal disposition in the sense that there is no information defectiveness. In this respect, the epistemological information is the same as epistemological knowledge which is equal to the ontological information and knowledge. The implication is that the lack of information defectiveness removes all elements that are noisy in the communication and decision-choice processes. The epistemological information is an exact replica of the ontological information which represents the ideal case. The lack of *information defectiveness* between the ontological and epistemological spaces also removes the concepts of possibility, possibility space and possibilistic reasoning on one hand and probability, probability space and probabilistic reasoning on the other hand in the knowledge production, as well as in the decision-choice outcomes with expectations, anticipations, risk, accidents and conditions of unsureness due to the fact that the perfect ontological information is mapped unto the perfect outcome space of knowledge.

Let us remember that in the ontological space there is no difference between information and knowledge. Knowledge is information and information is knowledge as input into the management of the command and control systems of decision-choice process of the dynamics of natural actual-potential polarities. The conditions of the ontological information is created in the ontological space where the epistemological information is a simple mimicry of the ontological information since their characteristic-signal dispositions are the same. The example of this type of epistemological space of information-knowledge process and knowledge viewed as an input into decision-choice system for command-control of socio-natural transformations is the classical mathematical space with full information and conditions of exactness. Under the conditions of exactness and quantitative completeness of the characteristic-signal disposition in the space of informing, knowing and learning, the paradigm of decoding, storage and analysis is the classical one. The corresponding space of analysis is a *non-stochastic* and *non-fuzzy information space* for decision-choice processes. The classical number and algebraic systems are example in this space where stochastic refers to the presence of limited information as seen in quantitative disposition, and fuzzy refers to the presence of vagueness as seen in qualitative disposition.

4.1.3.2 Exact and Incomplete Epistemological Information from the Ontological Space

Categorial differences between elements of ontological characteristic-signal disposition and epistemological characteristic-signal disposition arise from the imperfections of awareness as generated by acquaintance. It is the categorial differences at the level of epistemology from which noises in transmission tend to occur. The noises which arise from inexactness and incompleteness are epistemological but not ontological. The ontological space and activities are always in perfect states in terms of completeness, vagueness and exactness; they are the identities which present *what there is* (the collection of ontological varieties as it was presented in Chap. 3). The noises generate *uncertainties* in informing, knowing, and learning as well as in the knowledge production and decision-choice actions. The exact and incomplete acquaintance generates uncertainties due to *limitations on the number of ontological signals* encountered in the process of acquaintance. If S is the set of the number of signals encountered in the process of acquaintance and $S \subset \mathbb{S}$, then $(\#S) < (\#\mathbb{S})$ and $[(\#\mathbb{S}) - (\#S)] = \aleph > 0$ is the observational deficiency which is a rough measure of the categorial difference under conditions of exactness. The presence of categorial difference through the signal disposition implies that the epistemological set X is contained in the ontological characteristic set and hence $X \subset \mathbb{X}$ and $(\#X) < (\#\mathbb{X})$ with $[(\#\mathbb{X}) - (\#X)] = \aleph > 0$ as a numerical measure of observational deficiency. The pair $(X \otimes S)$ presents the epistemological characteristic-signal disposition which is the epistemological information.

The categorial difference presents uncertainties due to *quantitative limitation* in an *exact observational space*. It is this quantitative limitation of the epistemological characteristic-signal disposition that gives rise to *information defectiveness* of the quantitative type and the rise of probability, probability space and probabilistic reasoning in knowledge production, and decision-choice outcomes with probabilistic or stochastic expectations, anticipations, risk, accidents and conditions of unsureness due to the fact that the quantitative elements of the information defectiveness are mapped unto the outcome space of knowledge. This type of information defectiveness exists irrespective whether one approaches its analysis through the methodological classical probability or methodological personal probability. The epistemic approach to the construct of the probability space and the corresponding measure will depend on how one analytically conceptualizes uncertainty and information, and how information is seen as an input into the construct of the theory of knowledge where the results of the theory of knowledge become inputs into the theory of decision-choice systems for command-control actions of socio-natural transformations.

Under the conditions of exactness and quantitative limitation of the epistemic characteristic-signal disposition in the space of informing, knowing and learning, the paradigm of decoding, storage and analysis is the classical one. The corresponding space of analytical work, information processing and decision-choice action is called a *non-fuzzy stochastic information space* for knowledge production. There is the activity of knowledge production from information input since the conditions of equality between knowledge and information have been broken by the presence of defectiveness due to information limitation. The non-fuzzy stochastic space simply means that there is always quantitative unsureness associated with information transmission. The sureness of knowledge production is conditional on the sureness of the epistemological characteristic-signal disposition which then affects the sureness of outcomes in the management of the command-control process of decision-choice actions in the epistemological space.

4.1.3.3   Inexact and Complete Epistemological Information from the Ontological Space

On the other hand, if S is the set of the number of signals encountered in the process of acquaintance and $S \subseteq \mathbb{S} \subseteq \mathbf{S}$ then $(\#S) = (\#\mathbb{S})$, and $[(\#S) - (\#\mathbb{S})] = 0$. In this respect, there is no observational deficiency defined in terms of quantitative categorial difference under conditions of inexactness produced by vagueness through acquaintance. The presence of inexactness or vagueness, however, introduces a *qualitative categorial difference* and not a *quantitative categorial difference* in the process of acquaintance through the observational vagueness in the signal disposition, which implies that the epistemological characteristic set X as formed is equal to the ontological characteristic set and hence $X \subseteq \mathbb{X} \subseteq \mathbf{X}$ and $(\#X) = (\#\mathbb{X})$ with $[(\#X) - (\#\mathbb{X})] = 0$. The pair $(X \otimes S)$ representing the *epistemological characteristic-signal disposition* is the epistemological information and is taken to be equal to the *ontological characteristic-signal disposition* $(\mathbb{X} \otimes \mathbb{S})$. The presence of inexactness translates to the conditions that the equality of the epistemological signal disposition and the ontological signal disposition may not hold in that some elements in the acquaintance may be illusory, that is, $S \subseteq \mathbb{S} \not\subseteq S$, $(S \cap \mathbb{S} = S)$ and $(S \cup \mathbb{S} = \mathbb{S})$ with $(\#S) < (\#\mathbb{S})$ and hence $X \subseteq \mathbb{X} \not\subseteq X$ with $(\#X) < (\#\mathbb{X})$.

In this epistemic frame, there is a qualitative categorial difference that presents *fuzzy uncertainties* due to qualitative limitation in the *inexact observational space*. It is this qualitative limitation of the epistemological characteristic-signal disposition that gives rise to *information defectiveness* of the qualitative type and the rise of possibility, possibility space and possibilistic reasoning in the knowledge production, and decision-choice outcomes with fuzzy anticipations, risk, accidents and conditions of unsureness due to the fact that the qualitative elements of the information defectiveness are mapped unto the *inexact outcome space* of knowledge and then into the decision-choice system. This type of information defectiveness exists whenever there are conditions of vagueness and qualitative elements that require analytical subjectivity.

The analysis of the qualitative type of information defectiveness requires the use of methodological possibility, possibilistic reasoning and the *fuzzy paradigm of thought*. The epistemic approach to the construct of the possibility space and the corresponding possibilistic measure will depend on how one analytically conceptualizes vagueness, subjectivity, intentionality and information, and how information is seen as an input into the construct of the theory of knowledge, where the results of the theory of knowledge become inputs into the theory of decision-choice system for command-control of socio-natural transformations or dynamics of actual-potential polarities. Under the conditions of quantitative completeness and qualitative limitation of the epistemological characteristic-signal disposition in the space of informing, knowing and learning, the paradigm of decoding, storage and analysis is the fuzzy one. One may keep in mind that knowledge production is a process of information refinement to do away with noise that creates disparity between information and knowledge.

Under conditions of inexactness and quantitative completeness where vagueness characterizes the elements in the epistemic signal set S, subjectivity in the exactness of the signals received and their meaning become important in affecting the validity of S and hence $(\#S)$, and the true value of $(\#S - \#\mathbb{S}) = 0$ with the result that $(\#X - \#\mathbb{X}) < 0$. The vagueness in the acquaintance is a qualitative phenomenon and may be handled by introducing a possibility space and possibility measure through the degree to which the elements in S are exact. In other words, the acquaintance is seen in terms of perception which is subjective clarification of the elements of signal disposition. To correct the subjective clarification of the vagueness, a membership characteristic function $\mu(\cdot)$ that provides the degree of subjective exactness is attached to the set S in the form $\{(\mu_S(s),S)|s \in S, \mu_S(s) \in [0,1]\}$. The condition $\mu(\cdot) = 0$ implies that for a particular $s \in S$ the corresponding $x = x \notin \mathbb{X}$. In this way, even though $(s, x) \in (S \otimes X)$ but $(\#S - \#\mathbb{S}) = 0$.

The conditions are such that the vagueness in the epistemological signal disposition is mapped onto the epistemological characteristic disposition such that $\#(S \oplus X) < \#(\mathbb{S} \otimes \mathbb{X})$ which translates into $\sum_{x,s} \mu_{(X \otimes S)}(x,s) < \#(\mathbb{S} \otimes \mathbb{X})$, where $\mu_{(X \otimes S)}(x,s) \in [0,1]$ in the epistemological space, and hence $\{(\mu_S(s),S)|s \in S, \mu_S(s) \in [0,1]\} = \{(\mu_X(x),X)|x \in X, \mu_X(x) \in [0,1]\}$. The vagueness generates illusions in the perception of reality as seen in the ontological space. These illusions may be related to the epistemological perception of the characteristic-signal disposition of the results of the dynamics of actual-potential polarity in the ontological space, where there is actual-potential inter-transformability in that the actual may be transformed into potential and the potential may be transformed into actual, and hence the benefit becomes the cost and the cost becomes the benefit.

The corresponding space of analytical work, information processing and decision-choice action is called the *non-stochastic fuzzy information space* for knowledge production. The conditions of vagueness produced by acquaintance in the relational structure between the ontological and epistemological spaces are such that the equality between the members of the family of characteristic-signal dispositions in the ontological space and the members of the family of epistemological spaces are broken. As such, knowledge is not the same as information and there must be activities of knowledge production from information which must act as input, since the conditions of equality between knowledge and information have been destroyed by the presence of defectiveness due to vagueness in the information which contains knowledge. The non-stochastic fuzzy information space simply means that there is always qualitative unsureness associated with information transmission. The sureness of knowledge production is conditional on the qualitative sureness of the epistemological characteristic-signal disposition which then affects the qualitative sureness of outcomes in the management of the command-control process of decision-choice actions in the epistemological space.

4.1.3.4   Inexact and Incomplete Epistemological Information from the Ontological Space

It has been explained that categorial differences between elements of ontological characteristic-signal disposition and epistemological characteristic-signal disposition arise from the imperfections of awareness as generated by acquaintance. The categorial differences are made up of a *quantitative type* involving incompleteness and a *qualitative type* involving vagueness in the epistemological characteristic-signal disposition. It is the imperfections at the level of epistemology that give rise to noises in the source-destination transmission processes. The noises which arise from inexactness and incompleteness are therefore epistemological but not ontological. The noises generate *uncertainties* in informing, knowing, and learning as well as in the knowledge production and decision-choice actions. Information is no longer knowledge but a means to knowledge through a process. The incompleteness is associated with the quantity of the signal disposition and the inexactness is associated with the quality of the signal disposition. The incomplete acquaintance generates *stochastic uncertainties* due to cognitive *limitations on the number of ontological signals* encountered in the process of acquaintance in the epistemological space. The inexact acquaintance, on the other hand, generates *fuzzy uncertainties* due to cognitive *limitations on the quality of ontological signals* encountered in the process of acquaintance in the epistemological space.

The stochastic uncertainties and fuzzy uncertainties are the result of the relational structure of ontological characteristic-signal disposition and epistemological characteristic-signal disposition through the process of informing, knowing and learning. The presence of the stochastic and fuzzy uncertainties introduces defectiveness into the epistemological characteristic-signal disposition that forms the definitional foundation of epistemological information. The conditions for defining the epistemological information is no longer the same as those of ontological information. However, they can be abstracted where the ontological conditions constitute the primary category for defining the epistemological information and the epistemological conditions constitute the derived category for producing other definitions. The information that is obtained in the presence of quantitative and qualitative limitations on the epistemological characteristic-signal disposition is, here, referred to as *defective information* as compared to *perfect information* in the ontological space.

The analysis of the defective information due to the limitations of quality and quantity over the epistemological space is done by combining the methodological approaches for the analysis of exact and incomplete signal disposition and inexact and complete signal disposition. Since the qualitative and quantitative dispositions are of simultaneous occurrence the methods will need some adjustments. If S is the set of the number of signals encountered in the process of acquaintance and $S \subset \mathbb{S}$ then $(\#S) < (\#\mathbb{S})$ and $[(\#\mathbb{S}) - (\#S)] = \aleph > 0$ is the *observational deficiency* which is a rough measure of the *quantitative categorial difference* under conditions of exactness. The presence of quantitative categorial difference through the signal disposition implies that the epistemological set X is contained in the ontological characteristic set and hence $X \subset \mathbb{X}$ and $(\#X) < (\#\mathbb{X})$ with $[(\#\mathbb{X}) - (\#X)] = \aleph = [(\#\mathbb{S}) - (\#S)] < 0$. The quantitative categorial difference generates uncertainty in an exact relation between $\mathbb{X}$ and X that leads to the development of the space of probability where the exact relation between $\mathbb{X}$ and X may be viewed as an outcome. The presence of vagueness translates into the degree of sureness of acquaintance and observations of the ontological signal dispositions. The degree of sureness of each signal disposition has a membership attachment in the form $\mu_S(s) \in [0, 1]$.

The epistemic signal disposition may be written as $S = \{(s, \mu_S(s)) | s \in S, \mu_S(s) \in [0, 1]\}$ with the condition that the epistemological characteristic disposition is such that $\sum_{s \in S} \mu_S(s) < (\#X) \ll (\#\mathbb{X})$ in terms of cardinality of characteristics. The fuzzy-stochastic relative value $\eta = \frac{\sum_{s \in S} \mu_S(s)}{(\#S)} \ll \frac{(\#S)}{(\#\mathbb{X})}$ generates relative fuzzy-stochastic information where the pair $(X \otimes S)$ is transformed into $(X, \mu_X(x)) \otimes (S, \mu_S(s))$ which includes the assessment of degree of exactness in the comparative process of the ontological characteristic-signal disposition and epistemological characteristic-signal disposition, and presents a fuzzy-stochastic epistemological characteristic-signal disposition which is simply the *epistemological information* under conditions of qualitative inexactness in terms of vagueness and quantitative incompleteness in terms of volume. The quantitative-qualitative epistemological characteristic-signal disposition presents information defectiveness which generates uncertainties which are attributed to the presence of simultaneity of *quantitative-qualitative limitations* in an *incomplete-inexact observational space*. This incomplete-inexact observational space presents different types of noise in the source-destination transmission process. In this respect, the analytical structure must combine possibility and probability where the uncertainties are characterized as fuzzy-stochastic uncertainties due to the qualitative and quantitative limitations in the epistemological characteristic-signal disposition.

4.1.4   Noise and Deceptive Information in the Transmission Process

In Sect. 4.1.2, an analysis was provided on the *defective information* as producing noise in the source-destination transmission that connects the ontological space and the epistemological space of information activities. The defective information structure is composed of *stochastic information* due to quantitative limitations and *fuzzy information* due to qualitative limitations that affect the completeness and exactness of the signal disposition and cloud the characteristic disposition in such a way that the epistemological characteristic-signal disposition appears as defective for any variety relevant over the epistemological space. It was pointed out that the ontological information is perfect in the sense of completeness and the lack of defectiveness while the epistemological information is defective in one form or the other as the ontological characteristic-signal disposition is mapped onto the space of epistemological characteristic-signal disposition for any variety. The source-destination transmission of defective epistemological information, defined in terms of epistemological characteristic-signal disposition, over the epistemological space is further complicated by possible conditions of *deceptiveness*. The result of this complication involves the concept of *deceptive information structure*. The deceptive information structure in terms of deceptive characteristic-signal disposition introduces additional noises which amplify the existing noise due to defectiveness. In other words, there are noises due to defective information structure and noises due to deceptive information structure.

4.1.4.1   The Nature and Analytics of Deceptive Information Structure

The structure of defective information structure and the corresponding noises have been discussed. The space and manner in which deceptive information structure is generated are different. Defective information structure is a socio-natural phenomenon due to the limited cognitive capacity of cognitive agents in the process of forming the epistemological characteristic-signal disposition from the ontological characteristic-signal disposition. Its concepts and meaning are from ontology to epistemology. Examples are all scientific information whether they are empirically or axiomatically stated. The *deceptive information structure* is purely a social phenomenon that takes place over the epistemological space in the process of source-destination transmission of epistemological characteristic-signal disposition among cognitive objects. It originates from the source of cognitive agents to the destination of cognitive agents over the epistemological space with decision-choice intentionalities in accordance with some preference ordering. The defective information structure and deceptive information structures are distinguished by the intentionality of the source in the decision-choice systems over the epistemological space. The intentionality is always to create advantage. It is within this epistemic structure that semantic information theories are developed as an examination of the relationships between meanings as may be conceived in the source-destination duality [R12.4, R12.22], and mathematical theories of communication are developed to deal with the relationship between channels and the size of signal disposition [R8.1, R8.5, R20.11, R20.12].

Definition 4.1.4.1: Deceptive Information Structure
An epistemological characteristic-signal disposition is said to be a deceptive information if the nature of the characteristic disposition from the source (sender) is intentionally presented to create quantitative (dis-informative) and qualitative (mis-informative) limitations in a manner that brings limitations and vagueness in the signal disposition, such that the one-to-one correspondence between characteristics and signals is no longer valid at the destination (receiver).

The *deceptive information structure* is specifically associated with the collective decision-choice system of two or more cognitive agents, where conflicts of preferences tend to arise in cost-benefit configuration of social actions. It is made up of *disinformation* and *misinformation* substructures. It relates to language vagueness and representation, and ambiguities in transmission signals through information manipulations to change thoughts and direct preferences to the desires of the manipulators. The disinformation process is a strategy to empty the mind of the decision-choice agents of what is known to be knowledge and to create *cognitive emptiness* through the activities in the sureness-unsureness duality with relational continuum and unity. The misinformation process is a strategy to fill the cognitive emptiness of the decision-choice agents with information that may be made up of combined distorted signals to create a faked epistemological information from which knowledge is derived in support of particular decision-choice actions. Both disinformation and misinformation components of the deceptive information structure are associated with quality and quantity of the epistemological information given the defective information structure.

The disinformation process reduces the space of the quantitative disposition of the stochastic epistemological information. This affects the quantity of knowledge derived by the cognitive agent from further limitations on the epistemological characteristic-signal disposition. The result leads to the amplification of noise and the stochastic uncertainty as well as increasing the corresponding *stochastic risk* in the quality of knowledge and decision-choice actions in the social space, where quality of knowledge relates to degree of validity. The disinformation and misinformation are phenomena in the signal disposition that are associated with intentionality of the source. They are not phenomena in the characteristic disposition. The intentionality objective of disinformation and misinformation is to destroy the one-to-on relationships between the elements of the characteristic disposition and the elements of the signal disposition in order to affect the results of the knowledge-production process and decision-choice actions.

The misinformation reduces the quality of information and amplifies vagueness thus enlarging the domain of *fuzzy uncertainty* and the corresponding *fuzzy risk* in information processing and knowledge acquisition in the epistemological space of the decision-choice system. In this respect, it is useful to know that propaganda of

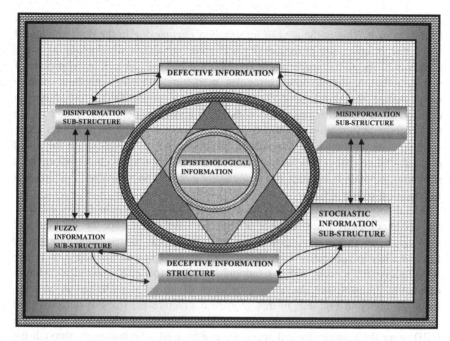

Fig. 4.1 A cognitive geometry of relational structures of defective and deceptive information substructures in defining the epistemological information relative to the ontological information

all kinds belongs to the domain of deceptive information, where false information is presented as genuine information. The general purpose of the deceptive information structure from the source is to create cognitive illusions in the minds of decision-choice agents through the development of information asymmetry between the source and the destination over the epistemological space in all the decision-choice systems. The relationships among these information substructures are presented in terms of cognitive geometry in Fig. 4.1. These relationships are translated into collective noise in the source-destination transmission mechanism. It is important to note that every message received at the destination is considered as a genuine information which must be processed into knowledge and then taken as an input into the decision-choice process. Its modalities of truth, necessity and possibility rest on the derived knowledge and the degree to which this knowledge can meet the conditions of methodological reductionism. The question of alethic or epistemic neutrality does not arise. However, the questions of source-destination intentionalities present themselves in acceptance or rejection of the information at the destination and the rejection or acceptance depends on the cognitive limitationality and logical limitativeness.

The deceptive information structure amplifies the penumbral regions of the decision-choice system and the associated systemic risk by affecting the outcomes of the derived knowledge by the destination. In a collective decision-choice system that involves conflicts and games, the deceptive information structure is under the manipulation from the source and by those who have resources to engineer and benefit from it. This is the area of *information engineering* in the epistemological space that is under the propagandists and game strategists who belong to every endeavor of the social space where conflicts are present (such as in conditions of war, negotiation, games, power struggles, marketing and many others) to continually change the cost-benefit distributions in the collective for advantage in decision-choice systems associated with every transformation. They prefer to use the deceptive information structure to influence collective decision outcomes under conditions of conflicts in preferences to create cost-benefit advantage in the social decision-choice space. The interactions of the defective and deceptive information structures generate a complex *fuzzy-stochastic information structure* in all areas of the decision-choice space.

The theory of information in concept, phenomenon, measurement and use over the epistemological space requires an enhanced definition over the definition of ontological information and generalized analytical tools of reasoning through the fuzzy-stochastic rationality with fuzzy laws of thought and mathematics in the fuzzy-stochastic system. The relational structure of epistemological source-destination transmission interference in terms of noises is presented as an epistemic geometry in Fig. 4.2. In this geometric structure the roles of defective and deceptive sub-structures in presenting the General Information Definition (GID) over the epistemological space are emphasized. The noises are through the intentionalities that lead to the manipulation of the elements in the characteristic-signal disposition such that the symmetric relation between the characteristic disposition and the signal disposition is no longer maintained. The structurally epistemic points to the idea is simply, that uncertainties and risks are the work of information through imperfections of the relational structure of the characteristic-signal disposition over the epistemological space. To what epistemic extent is it acceptable to use the measures of uncertainty and risks such as those involving probability and possibility to define the concept and phenomenon of information? Are these measures not information-concept dependent? If they are how can they be used to define the concept of information?

4.1.4.2 Analytics of the Complexity of Imperfect Information and Transformation Noise

The imperfect information space, I is made up of defective information space (I_D) and deceptive information space (I_Δ) such that $I = (I_D) \cup (I_\Delta)$ and $W = (I_D) \cap (I_\Delta) \neq \emptyset$. The nature of these information spaces are defined and established acquaintance conditions of the signal disposition $\mathbb{S}$ relative to the characteristic disposition $\mathbb{X}$. The defective information space is made up of subspace of limited information subspace (I_L) and vague information subspace (I_V) where $(I_D) = (I_L) \cup (I_V)$ and $O = (I_L) \cap (I_V) \neq \emptyset$. The deceptive information space is made up of disinformation space (I_Z) and misinformation space (I_M) with

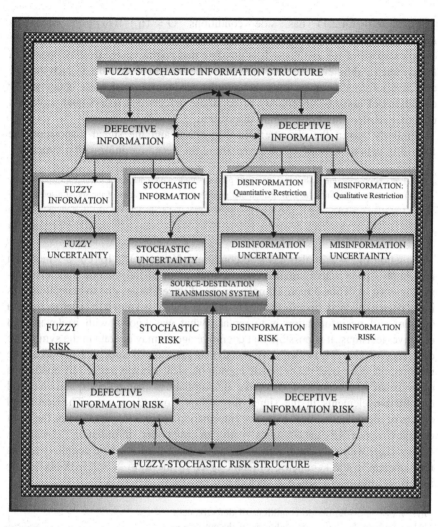

Fig. 4.2 An epistemic geometry of the relational elements of types of noise in General Information Definition (GID) over the epistemological space

$(I_\Delta) = (I_Z) \cup (I_M)$, and $X = (I_Z) \cap (I_M) \neq \emptyset$. Given the characteristic disposition and the signal disposition the following conditions may be observed about the defective information structure.

$$
\left.
\begin{aligned}
(I_L) &\Rightarrow (\#\mathbb{S} < \#\mathbb{X}) \text{ due to conditions of exact and incomplete acquaintance} \\
(I_V) \Rightarrow (\#\mathbb{S} > \#\mathbb{X}) &\text{ due to conditions of inexact representation and complete acquaintance} \\
(I_D) &= (I_L) \cup (I_V) \Rightarrow (\#\mathbb{S} \lll \#\mathbb{X}) \text{ due to conditions of limitation and vagueness}
\end{aligned}
\right\}
$$

$$(4.1.4.2.1)$$

Equation (4.1.4.2.1) has side conditions $O = (I_L) \cap (I_V) \neq \emptyset$ which is non-empty space where vagueness and incompleteness combine to define more complex space of uncertainties. It is possible that cognitive agents may operate in the symmetric difference of $S = (I_L) \nabla (I_V)$. The analytical structure is such that the defective information space over the complete epistemological space is defined by conditions of noise of uncertainty of cognitive defects that reflect qualitative and quantitative elements in the signal disposition.

Similarly, given the characteristic disposition and the signal disposition the following conditions may be observed about the deceptive information structure.

$(I_Z) \Rightarrow (\#\mathbb{S} < \#\mathbb{X})$ due to disinformation that produces underrepresentation of the signal disposition

$(I_M) \Rightarrow (\#\mathbb{S} > \# \mathbb{X})$ due to misinformation that produces overrepresentation of the signal disposition

$(I_\Delta) = (I_Z) \cup (I_M) \Rightarrow (\#\mathbb{S} \lll \#\mathbb{X})$ due to combined irrelevant underrepresentatin and overrepresentation

$$(4.1.4.2.2)$$

Equation (4.1.4.2.2) also has side conditions $X = ((I_Z) \cap (I_M)) \neq \emptyset$ which is a non-empty space where simultaneous operations of disinformation and misinformation exist to define a more complex space of uncertainties due to deceptions. As it has been explained, disinformation is to reduce what is known in order to create a vacuum and misinformation is to replace what was known with irrelevance to effective decisions. It is possible that cognitive agents may operate in the symmetric difference of $C = ((I_Z) \nabla (I_M))$. The analytical structure is such that the deceptive information space over the complete epistemological space is defined by conditions of noise of uncertainty of deceptive intentionalities that reflect qualitative and quantitative elements in the signal disposition from the source. The combined effects of the defective and deceptive information space is the creation of the most complex space of uncertainties with an imperfect information space $I = ((I_D) \cup (I_\Delta))$ with side conditions of $G = ((I_D) \cap (I_\Delta)) \neq \emptyset$ where cognitive agents may operate in the symmetric different space of the form $N = ((I_D) \nabla (I_\Delta))$. The imperfect information space is such that:

$(I_D) \Rightarrow (\#\mathbb{S} \lll \#\mathbb{X})$ due to conditions of defectiveness

$(I_\Delta) \Rightarrow (\#\mathbb{S} \lll \# \mathbb{X})$ due to conditions of deceptiveness

$(I) = (I_D) \cup (I_\Delta) \Rightarrow (\#\mathbb{S} \lll\lll \#\mathbb{X})$ due to conditions of defectiveness and deceptiveness

$$(4.1.4.2.3)$$

The symbol $(\lll)$ means substantially less than and $(\lll\lll)$ means extremely less than. It may be noted that when the signal disposition is less than the characteristic disposition for any phenomenon $\phi \in \Phi$ and any object $\omega \in \Omega$ then the true identity of $\omega \in \Omega$ may not be revealed to the cognitive agent(s). Let us keep in mind that the cognitive agents' activities involving informing, knowing and learning about all objects $\omega \in \Omega$ and all phenomena $\phi \in \Phi$ take place through the interactive effects of acquaintances and signal dispositions. The activities involving teaching, deciding and choosing take place through the interactive effects of the results of informing, knowing and learning and the behaviors of the cognitive agents. These are spaces in which cognitive agents operate under qualitative and quantitative dispositions in activities of informing, knowing, learning, teaching, deciding and choosing. Over the epistemological space, the characteristic disposition including intentionalities is hidden from the cognitive agents and must be discovered through the interactions between signal disposition and acquaintance. In other words, it is through the acquaintance-signal-disposition process that activities of informing, knowing, learning, and teaching, deciding and choosing take place.

The effects of the relational structure of the defective and deceptive elements in information as presented in Figs. 4.1 and 4.2 and Eqs. (4.1.4.2.1) and (4.1.4.2.2) in the construct of characteristic-signal disposition may be mapped onto the source-destination transition mechanism as channel noises that create uncertainties, risks and accidents which must derive their definitions and phenomena from the general definition of the concept and phenomenon of epistemological information. The definition of the concept of epistemological information must be such that a clear distinction is placed on the ontology as establishing the *primary category of varieties* and the epistemology as establishing the *derived categories of varieties*. In other words, the existence of the epistemological categorial varieties is a direct or indirect derivative from the ontological categorial varieties. In respect of this, the *epistemological General Information Definition* (EGID) must be a derivative of the *ontological General Information Definition* (OGID). The system of noises is a property only of the epistemic space where imperfections are generated by the epistemic and cognitive processes in the structure of informing, knowing, learning and teaching.

The complete noise system in the source-destination transmission mechanism is shown in terms of epistemic geometry in Fig. 4.3. The General Information Definition (GID) over the epistemological space becomes the *primary category* of the information definition on the basis of which all other definitions are *derivatives* in terms of epistemological characteristic-signal dispositions. The nature of such specific derivative definitions of information are driven by specific needs in areas of application such as computer information, economic information, empirical information, axiomatic information and others, all of which relate to characteristic-signal disposition in terms of distribution of specific varieties. On the basis of this epistemological characteristic-signal disposition that displays epistemological varieties, all other definitions of the concept and phenomenon of information can be shown to contain *epistemological varieties* if the concept and phenomenon of information are to be analytically useful. These epistemological varieties may deviate from their corresponding ontological varieties. The verification of the correspondence between any ontological variety and epistemological variety is through the knowledge-production process under the combined interactions of methodological constructionism-reductionism duality, given methodological nominalism. The inputs in the epistemological space for research are ontological signal dispositions and the outputs are epistemological signal dispositions in the process of discovering varieties in comparative analysis of ontological characteristic dispositions and epistemological characteristic dispositions [R4.10, R4.13].

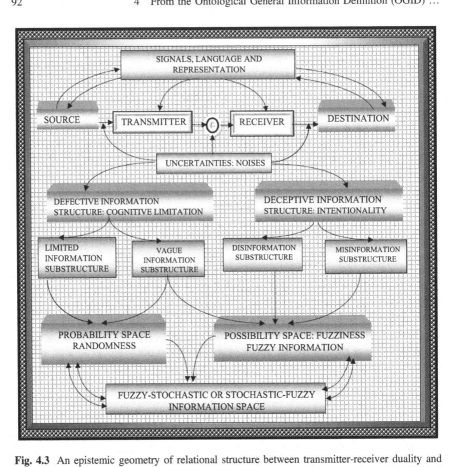

Fig. 4.3 An epistemic geometry of relational structure between transmitter-receiver duality and source-destination duality under conditions of non-standard information over the epistemological space

It has been pointed out that the question of the epistemic modality of truth, necessity, possibility and probability in the meaning is not a useful way of assessing the effects of deceptive information structure in the epistemological information. The most important thing is the *intentionalities* in the source-destination mechanism or the activities in the sender-receiver duality with relational continuum and unity. The point here is that every action in the decision-choice system or knowledge production system has intentionality, whether direct, indirect, known or unknown to the destination agents, connected to information which is defined in terms of conditions of varieties, and may have differential impacts on transformational direction and outcomes in epistemological actual-potential polarities through the activities in the set of dualities, where each duality exists with relational continuum and unity. The noises due to deceptive information activities may reflect the sender-receiver intentionalities of the nature of the preference of distribution of outcomes of epistemological varieties. It must be clear, as presented in this analytics, that uncertainties are noises which will also affect the source-destination transmission capacity. It does not seem helpful to speak of *false information* and *true information* without reference to intentionalities of messages. The concept of true-false information will be taken up in the next chapter.

Epistemological information involves systemic risk in knowledge production and decision-choice actions in all areas, where the systemic risk may also be amplified by the limitations of the information-processing capacities of individual and collective cognitive agents acting in knowledge-production and decision-choice systems. The uncertainty-risk system that has been presented here and elsewhere, suggests that the definition of information as applied in the epistemological space must take account of the uncertainty-risk profile that is a derivative of the phenomenon of epistemological information and not the other way around. The probabilistic reasoning as motivated by stochastic uncertainty does not offer a channel for defining the concept of epistemological information neither does the possibilistic reasoning as motivated by fuzzy uncertainties. The definition of the concept of information must make allowance for defective and deceptive elements of epistemic processes in which information acquires intrinsic and use values. The defective-deceptive process generates a complex fuzzy-stochastic information structure in the epistemological space such that its processing and use in decision-choice systems require enhanced analytical tools of reasoning through the fuzzy paradigm for dealing the fuzzy-stochastic rationality with its laws of thought and mathematics in the fuzzy-stochastic topological space that must deal with conditions of quality-quantity dualities within subjective-objective principles of knowing, learning and teaching, where epistemic claims must meet the fuzzy conditionality.

The definition of the concept of epistemological information must acknowledge these elements of analytical necessities. The analytical necessities include quality, quantity, subjectivity and objectivity, where intentionalities must be related to the communication path of the contents of information (characteristic disposition) through the signal disposition. Under these conditions, any theory on objectivity of information in the epistemological space must be such as to admit of its opposite of subjectivity under the principle of opposites. The subjective phenomenon generates outcomes in the relational structure in the truthfulness-untruthfulness duality with relation continuum and unity. In this frame truthfulness and falsity exist as simultaneous property in all claims.

4.2 The Summary of the General Information Definition (GID) in the Epistemological Space

On the basis of the discussions and the definition of information it is useful to state the basic requirements of any information definition that may be derived from the GID. From these requirements it will be shown that any definition of information is a derivative of the GID in the sense that it projects distinction, difference and similarity through a distribution of varieties where each variety is established by a characteristic-signal disposition.

Let Θ be a collection of all possible source-destination objects with a generic element $\phi \in \Theta$ in the epistemological space for the sending-receiving information process. It may be noted that $\subset \Omega$ and the elements in Θ have special cognitive characteristics that allow them to be in both Θ and Ω but not every element $\omega \in \Omega$ belongs to Θ, which is a special subset of the universal object set Ω. In other words Θ is a collection of varieties. Let $\mathbb{V}_{\ell j}$ be an information structure from source $\phi_\ell \in \Theta$ to destination $\phi_j \in \Theta$ in the epistemological space, then $\mathbb{V}_{\ell j}$ is an instance of the GID of the concept of epistemological information if:

1. $\mathbb{V}_{\ell j}$ consists of a maximal characteristic set X_ℓ where $(\#X_\ell) \geq 1$ with a corresponding maximal signal set S_ℓ and $(\#S_\ell) = (\#X_\ell) \geq 1$ with $\#(X_\ell \otimes S_\ell) \geq 1$ as the maximal epistemological characteristic-signal disposition which establishes categorial varieties in the epistemological space.

2. The all characteristic-signal dispositions of the form $Z_\ell = (X_\ell \otimes S_\ell)$ are well-formed about the varieties with source-destination intentionalities.

3. Each well-formed characteristic-signal disposition, $Z_\ell = (X_\ell \otimes S_\ell)$ from $\phi_\ell \in \Theta$ to $\phi_j \in \Theta$ is meaningful to all $\omega_j \in \Omega$ in the epistemological space with equal or differential cognitive capacities for information activities and interpretations of intentionalities of the well-formed characteristic-signal disposition.

4. The well-formed characteristic-signal dispositions of the form $Z_\ell = (X_\ell \otimes S_\ell)$ from the source $\phi_\ell \in \Theta$ to the destination $\phi_j \in \Theta$ contain objective-subjective, qualitative-quantitative, defective-deceptive, complete-incomplete and exact-inexact elements in transmission, such that there is a set of noises N that gives rise to fuzzy-stochastic uncertainties, accidents and risks in their use.

5. If $Z_j = (X_j \otimes S_j)$ is the received characteristic-signal disposition by $\phi_j \in \Theta$, then $Z_j \gtrless Z_\ell$ depending on the nature of the noise-set N that establishes deferential source-destination interpretation and understanding of the structure of $\mathbb{V}_{\ell j}$.

6. There are conditions such that $\#X_j \gtrless \#S_j$ which generate an error $|\epsilon| = |\#X_j - \#S_j| > 0$ in the recipient system in such a way that $\#X_j \gtrless \#X_\ell$ and the error is verified through the knowledge-processing function $F : S_j \to X_\ell n$.

Note

The space of acquaintance and observation is such that disparities may occur between the epistemological characteristic disposition and the epistemological signal disposition for any given variety. The result is that the following conditions may be established.

$$\text{If} \begin{cases} (\#X_j - \#S) > 0 \Rightarrow \text{observational illussion} \\ (\#X_j - \#S) < 0 \Rightarrow \text{observational difficiency} \\ (\#X_j - \#S) = 0 \Rightarrow \text{perfect observation} \end{cases} \Bigg\} \text{in the space of acquaintance}$$

$$(4.2.1)$$

The epistemological characteristic-signal disposition is the derived category of the information definition that establishes the phenomenon of information relative to the primary category of ontological characteristic-signal disposition that forms the basis of the General Information Definition. It is argued that all other types of the epistemological information definition are obtained by methodological constructionism and may be shown to be logical derivatives of the ontological information as the primary category by methodological reductionism. Similarly, all categories of the information definition over the epistemological space may be arrived at from the ontological information by methodological constructionism. The channel of transmission of ontological information is argued to be *noiseless* where information imperfections are shown to arise only in the epistemological characteristic-signal disposition through the logical movement from the ontological space to the epistemological space. The general information definition in the ontological space is what has been called *ontological information* in this monograph. It is the concept of perfect information under the General Information Definition. The general information definition in the epistemological space is what has been called *epistemological information* in this monograph. The channel of transmission of epistemological information is argued to contain different types of noises which distinguish it from the ontological information with *noiseless* transmission channels. The definition of the concept of epistemological information is imperfect information under the General Information Definition (GID). The condition (6) above will be revisited when we speak of *data* in terms of its definition, meaning and utility. It may be kept in mind that the epistemological information is contained in the ontological space. It is very small in relative size.

The ontological information defines the ontological space by fixing the morphology of the elements in the universal object set where these ontological objects appear as varieties and categorial varieties that present differences and similarities in terms of characteristics. The study of the morphology of these elements is the study of the ontological information as captured by characteristic-signal dispositions that specify the varieties and categories of elements of *what there is* (the actual) and *what would be* (the potential) within the dynamics of natural actual-potential polarity, where some old varieties may be destroyed and new varieties may be created under the conditions of continual differentiation by socio-natural forces. The study of ontological objects is the study of ontological information. The epistemological information, on the other hand, defines the epistemological space by fixing the structure and form of the elements in the epistemological object set, where the epistemological objects appear as varieties and categorial varieties that present differences and similarities in terms of characteristics as abstracted from the ontological structures. The epistemic and decision-choice activities in the epistemological space are attempts to understand the morphology of elements in the ontological space through the creation of derivatives from the ontological information and the creation of epistemological characteristic-signal dispositions that specify the derived varieties and categories of elements of *what there is* (the actual) and *what would be* (the potential) within the dynamics of socio-natural actual-potential polarity. The knowing—learning activities about the ontological objects are the creation and analysis of epistemological information through the methodological duality of constructionism and reductionism with a designed paradigm of thought under methodological nominalism and representation of the signal dispositions with their corresponding characteristic dispositions.

4.2.1 Reflections on the General Information Definition (GID)

To conclude the discussion on the general information definition over the epistemological space, it is useful to place emphases on some ideas about source-destination information processes within conflict-cooperation duality regarding intentionalities and subjective aspects of information. These subjective aspects of information are intimately related to the ontological elements operating in epistemological space with intentionality and decision-choice capacities. There are three types of categories of information that have been identified in this monograph. The first category of information connects general ontological objects in the ontological space for informing, knowing and learning for adaptation and continual transformative actions in the ontological space. This is subjective-objective information relation in the ontological space. The second category of information is a connector between the ontological space and the epistemological space; thus between ontological objects and epistemological objects for the activities of informing, knowing, learning and decision-choice actions in the universal environment for transformation with intentionalities. This is objective-subjective information relation between ontological objects and epistemological objects in the general ontological space. The third category of information connects objects in the epistemological space for informing, knowing, learning and decision-choice actions with intentionalities of conflict and cooperation which are established by preferences of objects that belong to both ontological space and epistemological space. This is subjective-objective information relation of communications which are established among epistemological objects in the epistemological space. The general transmission structure is from ontological objects to ontological objects, from ontological objects to epistemological objects as a subset of ontological objects and from epistemological objects to epistemological object. The ontology—ontology and epistemology-epistemology are information mappings into themselves.

The theoretical structure begins with the postulate of characteristic-signal existence where the universe is seen as composed of a set of objects and characteristics. Every element in the universal objects set is composed of a set of characteristics that establishes its identity, distinction, difference and similarity in terms of varieties. This is the concept of *characteristic disposition*. In other words, the universe is a collection of varieties. Each characteristic has capacity to send a signal for its presence. Supporting the set of characteristics is a set of signals that corresponds to the set of characteristics. This is the concept of *signal disposition*. The characteristic (attribute) signals create conditions for awareness and inter-awareness among the universal objects in terms of relations between the source objects and recipient objects in ontological space. It is through these concepts of characteristic disposition and signal disposition that the *general information definition* of the concept and phenomenon of information are established as *characteristic-signal disposition* where a distinction is placed between ontological characteristic-signal disposition and epistemological signal disposition. The *ontological characteristic-signal disposition* as defining the *ontological information* is taken as the *primary category* of the general information definition by identity from which the *epistemological characteristic-signal disposition* as defining the epistemological information is a *derivative* by methodological constructionism. From the General Information Definition over the epistemological space, other definitions over the epistemological space may be abstracted where the common link is always characteristic-signal disposition that establishes varieties and categorial varieties. This approach allows one to speak meaningfully about specific information concepts with adjectives such as economic information, financial information, electronic information and many others. The adjectives are used under the definitional principle of the characteristic-signal disposition to create information varieties in the epistemological space.

   The first category of information is established when the awareness is among the ontological objects and the environment that they mutually establish. The second category is established when the awareness is between the ontological objects and epistemological objects. The third category is established when the awareness is either between or among epistemological objects. One thing that must be understood is that every object in the universal object set is both an information source (sender) and an information destination (recipient) in the sense that it sends and receives information through signals. It is here that the source-destination process and mechanism are defined in the sender-recipient duality with intentionalities and noises. In this way, the sender-recipient modules establish objective-subjective relationships that are defined by information flows and subjectively interpreted by the recipients for the knowing and decision-choice actions in a continual dynamics of social polarities.

   The General Information Definition (GID) as offered in this monograph divides the concept of information into two inter-supportive sub-concepts of *characteristic-based information* and *relation-based information* with the subjective-objective duality as seen in the environment of qualitative and quantitative dispositions. It is here that the concept of data and its relation to information finds epistemic meaning. The characteristic-based information is defined by *attributes* of objects, states, processes and events that exist independently of awareness of objects in the universal object space, where the characteristics are defined in quality-quantity space where the cognitive activities may attach meanings and interpretations. The relation-based information defines the characteristic signals between the source objects and recipient objects where naming, meaning and interpretations are attached to the characteristic (attribute) signals. The naming, meaning and

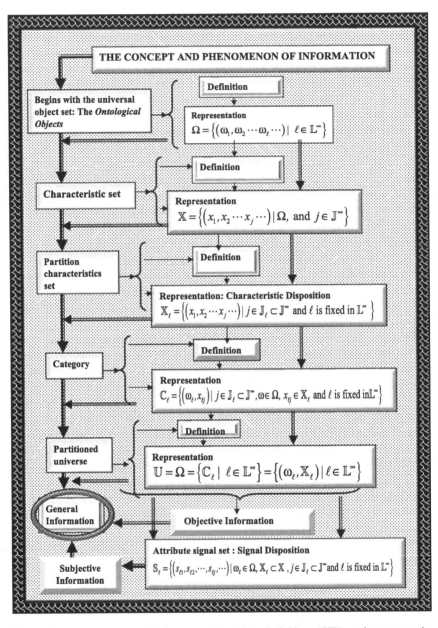

Fig. 4.4 Conceptual system of General Information Definition (GID) and representation characteristic-signal disposition in any given language

interpretation of received signals require capacities for processing the mechanism of the received characteristic signals. The meaning and interpretation that result from the processing mechanism of characteristic signals require awareness capabilities of recipient objects. As such, information relations among objects, states and processes are subjective, defined in signals with qualitative disposition whose measure in terms of quantitative disposition may help to establish the degree of information exactness and completeness as well as their epistemic representations. It is here that methodological nominalism in terms of the system of languages and grammar become useful to allow one to speak of semantic information which has meaning only over the epistemological space and between epistemological objects. In this discussion, the concern is on the content and phenomenon and not simply on the object-object communications in terms of human-human interactions, human-machine interactions or machine-machine interactions over the epistemological space.

The view of the organic concept of information as two interrelated sub-concepts of objective and subjective puts information into the heart of the philosophical problem of *what there is* (the objective) in either physics, cosmology, chemistry or social science and how *what there is* (the *it*) can be known (subjective). This approach is an objective-subjective theory of information that forms the foundation for the analysis and explanation of scientific theory in general and specific theories that may be encountered in different areas of knowledge production such as economics. The generality of the definition by the way of characteristic-signal disposition is such that it is neutral to all subject areas of human endeavor. The specific introduction of the concepts of qualitative and quantitative dispositions allows one to introduce the *fuzzy paradigm* and corresponding *fuzzy rationality* as a generalized concept of rationality into the process of informing, knowing, and learning into the informing-knowing process [R2, R4, R5, R6]. This will help to answer some fundamental questions such as: *what does the knower know? How does the knower know he or she knows? How do other cognitive agents believe in what is known? What does the learner learn, how does the learner believe that he or she has learned and how do others believe that the leaner has actually learned?* The answers to these questions rest on the characteristic-signal dispositions and their processing to abstract varieties in the epistemological space. The conceptual system of definitions and different types of information representation is shown in Fig. 4.4. This general information definition provides the set of static conditions of varieties and categorial varieties. It is the info-statics that relates to the history of socio-natural objects at any point of time. These set of static conditions become the initial conditions for the study of dynamics of varieties and categorial varieties, where the dynamics relate to information production.

Chapter 5
Information, Varieties, Categories, Ordering, and Socio-natural Transmissions Over the Ontological and Epistemological Spaces

It may be recollected that the GID in both ontological and epistemological spaces is variety-based where the varieties are captured by the distribution of characteristic disposition and transmitted by the distribution of signal disposition. The GID is characteristic-signal disposition that relates the concept and phenomenon of information to varieties. The ontological GID is perfect in the sense of being exact and complete in characteristic-signal disposition, and is transmitted in noiseless channels in such a way that there are no uncertainties, risk and others in the ontological space. It is the distribution of identities in terms of the primary category of knowing through acquaintance that provides the essentiality of the distribution of the characteristic-signal dispositions of varieties. The utility of ontological information is source-destination symmetric in cost-benefit configurations in the dynamics of natural actual-potential polarities. The epistemological GID is imperfect in the sense of being inexact and incomplete and sometimes deceptive in characteristic-signal dispositions, and is transmitted over noisy channels in such a way that there are uncertainties, risk and others in the epistemological space.

The utility of epistemological information may be seen in terms of either source-destination symmetric or asymmetric in cost-benefit configurations in the dynamics of social actual-potential polarities which relate to transformation of varieties and categorial varieties which shall be discussed in the volume on *info-dynamics*. Let us keep in mind that the dynamics of socio-natural actual-potential polarities involve transformations that are due to the production of necessary and sufficient conditions for qualitative changes in terms of characteristics that define inter-categorial varieties, and quantitative changes in terms of characteristics that define intra-categorial varieties. These characteristics which place distinction, differences and commonness on both ontological and epistemological objects, states and processes constitute the general information in the processes of informing, knowing and learning to allow the understanding of the necessary and sufficient conditions of transformations and existence of intra-categorial and inter-categorial differences. The creation of the necessary and sufficient conditions is the work of the interactions among matter, energy and information that destroys old varieties of

© Springer International Publishing AG 2018 101
K.K. Dompere, *The Theory of Info-Statics: Conceptual Foundations of Information and Knowledge*, Studies in Systems, Decision and Control 112, DOI 10.1007/978-3-319-61639-1_5

objects, states and processes and create new varieties with continual updating of the stock of information in terms of past-present-future varieties as established by the distribution of the characteristic-signal dispositions.

5.1 Information, Variety and Categories

Over the epistemological space, the varieties and categorial varieties present conditions of ranking and ordering holding qualitative disposition constant. This is intra-categorial ranking and ordering where the comparison is done within a category. Similarly, the varieties and categorial varieties present conditions of ranking and ordering of objects in the domain of qualitative disposition at a constant quantitative disposition. This is inter-categorial ranking and ordering where the comparison is done among categories. These inter-categorial and intra-categorial types of ranking and ordering are impossible without information on varieties as defined by distribution of characteristic-signal dispositions which must be processed into knowledge as input into the ranking-ordering decisions. The necessary and sufficient conditions for ranking and ordering in the qualitative-quantitative dispositions are established by varieties as defined by the distribution of characteristic dispositions and revealed by the distribution of the signal disposition. Therefore, and generally any change or transformation in both the ontological and epistemological spaces is about a change or transformation of varieties and hence a change of the characteristic-signal dispositions that establish the processes of informing, knowing, learning and the decision-choice actions underlie the forces of transformations. In this respect, all engineering forms are transformations where old varieties are destroyed and new varieties are created, and where the characteristic-signal dispositions of old varieties remain in an indestructible domain of historic existence and the characteristic-signal dispositions of the new varieties increase the stock of information in stock-flow disequilibrium processes in both the ontological and epistemological spaces.

Over the epistemological space, the knowledge is obtained by processing the epistemological signal disposition. The results of the knowledge are used as inputs into the ordering-ranking decision and the choice actions on the basis of specified rationality, where the outcomes of the decision-choice processes become a growth of information to be processed into a new level of the *stock of knowledge* which is consistent with the level of the stock of information. When the new addition of knowledge is transmitted from a source to a destination, it becomes information that must be processed into a secondary knowledge at the destination due to the existence of noise and especially its deceptive component. In pure epistemic sense, it is useful to observe that knowledge is a derivative of information and hence it is not information at the source. It becomes information when it is transmitted from a source to a destination due to the noises. It, however, becomes knowledge under noiseless transmission. This is where the concept of the information-decision-interactive process takes its analytical strength in systems of simplicities and complexities. It is also here that ordering and ranking of elements and categories are determined by the information-variety process in the socio-natural spaces where knowledge is obtained by processing the signal disposition of categorial varieties of objects, states and processes. In the total epistemic process, the signal dispositions of categorial varieties are indirect properties of matter through the characteristic dispositions of categorial varieties. To deny matter-information connectivity in terms of primary-derivative processes is to deny the very nature of *what there is* and *what there is not* as well as to deny the existence of identities, varieties and differentiations associated with socio-natural transformations. In fact, our number system and any language have no meaning without categorial variety and its existence. The usefulness of our number system and all kinds of languages including mathematics demands conditions of categorial variety whose difference, similarities, commonness and other are the foundations of abstract mathematics such as set theory, group theory, category theory, nations, ethnicity, tribes and others.

In terms of intra-categorial and inter-categorial ranking, knowledge is a derivative from information and not otherwise in the sending-receiving process over the epistemological space. It is the knowledge as a processed information which becomes input into the decision-choice system to manage command-control actions of socio-natural transformations. Knowledge is, thus, that part of information which has gone into an analytical processing machine to become a useful input into decision-choice systems. It is information and not knowledge which is encoded, decoded, collected, stored and transformed into knowledge with the use of an acceptable paradigm of thought in the epistemological space. It may be kept in mind that when knowledge is sent from a source as a primary knowledge, it will appear as information in any destination, where the signal-disposition part of the characteristic-signal disposition is the subject of encoding-decoding actions in the communication process, and hence the primary knowledge from the source as an information to the destination must be processed into a secondary knowledge. In other words, over the epistemological space the source-destination communication process entails primary and secondary knowledge. To the destination agent, the primary knowledge is information and the secondary knowledge is the accepted knowledge. This secondary knowledge at the destination may deviate from the primary knowledge from the source. For example, learning with supervision such as teacher-student structure may be seen as a source-destination communication structure. This disparity between primary knowledge and secondary knowledge in the source-destination process brings some essential problems encounter in the teaching-learning duality into focus. The importance of information is its role as an input into the knowledge-production process. Its importance to the decision-choice system which functions through knowledge is indirect. Thus, there is information as input into knowledge-production the results of which become potential inputs into decision-choice systems whose outcomes become information flows to up-date the information stock. It is this interconnectedness of information, knowledge and decision-choice actions that guarantees the phenomenon of stock-flow disequilibria in the laws of *info-dynamics* which are seen in terms of interdependent relational structure of the past-present-future information under the general principles of continuum and unity for all varieties and categorial varieties.

5.1.1 The Source-Destination Relationship Over the Epistemological Space

Over the epistemological space, the cognitive process requires an establishment of clear distinction between informing and knowing and how they relate to learning. It is the need for this distinction which raises a number of important questions as a prelude to the decision-choice process in order to affect the quality of the management of command-control actions in socio-natural transformations. At the level of informing, the following questions arise in the epistemological space. Who is informing? This is the source-transmission or *sender-transmission question*. Who is informed? This is the destination-transmission or *receiver-transmission problem*. What is informed? This is a *content (message)-transmission problem* which involves quantitative-qualitative dispositions that in turn involve the *source-destination credibility problem* in the space of the deceptive information structure. How is one informed? This is the *transmission-efficiency problem* between the source and destination. What are the reasons for informing and what are the reasons for receiving the informing content? This is the *source-destination intentionality problem* of the communication process. How does one become aware that one is informed? This is the *destination-validation problem* of the conditions of informing relative to intentionalities. The answers to these questions lead to the development of the *general theory of information* which must include concept, definition, content, measurement, transmission and utility of the phenomenon of information. Interestingly, the available literature on information theory fails to answer most of these questions. Most of the literature on information is devoted to the communication aspects of semantic and measurement variables without explicit discussion on what is the possible meaning dispute about or what is being measured.

At the level of knowing, one is confronted with a corresponding set of questions. Who is the knower? This is the *destination-source-action problem*. What does the knower know? This is the *destination-content problem* which involves qualitative-quantitative dispositions of the result of knowing. What process does the knower get to know? This is the *destination-transmission problem* of the process of knowing. How does the knower know that he or she knows? This is the *destination-cognition-test problem* of what and how. What are the reasons for knowing? This is the *destination-intentionality problem* involving the knowing process. The answers to these questions and related ones have led to the development of forms of the *theory of knowledge*. At the level of learning activities, there is also a set of important questions to which the answers must be sought. Who is the learner? This is a *learning-source problem* of the learning action. Given the learner, what does the learner learn? This is the *learning-content problem* that involves quantitative-qualitative dispositions of activities of learning. How does the leaner learn? This is the *transmission problem* of content in cognition. What are the reasons (motivation) to learn? This is the *intentionality problem* in the learning process. How does the learner validate the result of learning? This is the test of the *mimicking-success problem*. What improvement can the learner make to the content of what is learned? This is the *epistemic-creativity problem*. The search for the answers to these questions has led to the development of forms of the *theory of learning*. In all epistemic frames, there is a close and inseparable relationship between the *theory of knowledge* and the *theory of learning* and this relationship is well connected to the *theory of informing*. In a simple form, the interconnected relationships of the theory of knowledge, the theory of learning and the theory of information then suggest that the theory of information does not belong to any area of knowing, such as mathematics. To claim that the theory of information is an area-specific will constrain the development of the general theory of information and its relationship to matter, energy and existential history, where the theory of information must account for the existence of qualitative and quantitative dispositions as time progresses.

It is analytically useful to point out the complex nature of source-destination and problem-solution dualities. In source is a destination and in a problem is a solution in relational continua and unity. Every source has a destination support and every destination has a source support. Every problem has a solution that resides in the problem and every solution has a problem that resides in the solution in the logic of categorial conversions. Informing, knowing, learning and teaching form a relational continuum and epistemic unity over the epistemological space. The results of informing become inputs into knowing, the results of knowing become inputs into teaching and the results of teaching become inputs into learning, while the results of learning creativity in enhancing the contents of learning become inputs into informing. In the same relational process, the theory of information, the theory of knowledge, and the theory of learning are not only connected but are unified over the epistemological space. In this interconnectedness and epistemic continuum and unification of informing, knowing, teaching and learning, a set of element in the defective information is carried to knowledge and learning, the ordering-ranking process of categorial varieties, and then to the knowledge-decision system in terms of uncertainties and risks for unintended outcomes. The deficiency in the decision-choice system thus arise from the information processes about the varieties and categories the knowledge of which becomes an input into ranking and ordering in the decision-choice space that affects the directions of relevant transformations.

5.1.2 Some Reflections on the General Information Definition and the General Theory of Information

It is analytically useful to have a short reflection on the general theory of information and its relational structure with the General Information Definition and its relationship to the general theory of information and what it seeks to accomplish within the general cognitive project of the epistemic activities of cognitive agents. In other words, what is the epistemic epicenter of the general theory of information as the primary category to abstract derived epicenters of sub-theories of information in terms of content and communication? The epicenter of the general theory of information is on the information content, information transmission, and information communication in time and over time. The general theory of information is about the content-communication process that defines the content phenomenon and the transmission phenomenon which together establish a communication system. The communication system involves content identification and content transmission in the processes of informing and knowing, and the development of knowledge in time and over time. The content identification is associated with characteristic disposition while the content transmission is associated with signal disposition. The characteristic disposition and the signal disposition combine into characteristic-signal disposition to establish information on varieties and categorial varieties.

At the source, the content identification is simply on the characteristic disposition of what variety is expected to be received at the destination through the transmission of the corresponding signal disposition. At the destination, the content identification is through the interpretation of the signal dispositions which are subjective interpretations that reveal the underlying characteristic dispositions associated with varieties. Viewed in this epistemic way, any system or subsystem may be defined as interrelated sets of intra-categorial relations of varieties and inter-categorial relations of categories of existence. Both intra-categorial and inter-categorial relations are information defined. In other words, systems are nothing but information relations of varieties and categorial varieties where the distribution of the characteristic-signal relations defines the distribution of varieties in both ontological space and epistemic space.

The relations and their interactive conditions of varieties and categorial varieties create an infinite set of enveloping paths of problem-solution processes. The structure of the problem-solution activities is that the intra-relations and inter-relations generate internal and external conflicts and problems at each solution round. When each problem is solved, a new problem arises from the solution. The nature of the new problem will depend on how the previous problem was solved. Each problem presents information and the corresponding solution presents information in varieties which are defined by characteristic-signal dispositions. Each path of the problem-solution activity is an information stock-flow process on varieties and categorial varieties to generate an expanding information set. Such an information set presents itself as a nested set in the time domain for each variety and each categorial variety. This analytical framework, where the concept of info-statics is extended to the time domain will be taken up in the next monograph reserved for the development of the theory of info-dynamics [R17.18]. The *general theory of information*, therefore, is seen as composing of two aggregate sub-theories of the *theory of info-statics* and the *theory of info-dynamics*. The theory of info-statics establishes the initial conditions for the development of the theory of info-dynamics. The theory of info-statics deals with the contents and transmission of information about varieties and categorial varieties at a point in time and hence provides analytical conditions of available information stock. It is thus, devoted to the study of the static conditions of ontological and epistemological varieties and categorial varieties over the epistemological space. The static conditions relate to definition, identification problem of similarities and differences of varieties, and categorial varieties at specified time points. The theory of info-dynamics deals with transformation of ontological and epistemological varieties and categorial varieties due to changing contents, and transmission of information over time and hence provides analytical conditions of the information stock-flow process. In its bare essentials, it is the study of ontological-epistemological information stock-flow processes. The information stock-flow conceptual system as relations of varieties and categorial varieties leads to the question of what is the value of information to which we now turn our attention.

5.2   The Value of Information

It has been argued in this monograph that matter, energy and information reside in relational continuum and unity. In fact, it has been argued that information is the property of matter and energy. This monograph does not accept the notion that information is information and not matter. One can also pontificate that information is information and not energy while energy is energy and not matter. This line of thought leads one to the blind alley of epistemic vagueness and not into the intellectual house of ignorance reduction. The relational continuum and unity of matter, energy and information are such that matter has value, energy has value and hence it must be expected that information has value also. What then is the value of information? In the above sections, it was postulated that the value of matter in the universal system is *existence* as recognition of varieties and categorial varieties *and transformations of varieties* as the primary category of value of being where all other values are derivatives thereof (all human artifacts are derived categorial varieties). The value of energy in the universal system is *power and work* as the primary category of value of being where all other energy values are categorial derivatives. The value of information in the universal system of states and processes is knowledge and decision as the primary category of value of being where all other information values are categorial derivatives from the primary value. The use of information in the universal system demands the conditions of identity recognition of varieties through multi-lateral and bilateral relations. All three categorial values are in relational continuum and unity in existence, where it becomes necessary to distinguish between *use value* and *intrinsic value* and how they are related to social-natural command-control actions in the universal space of continual trans-formation. The use value and the intrinsic value are made possible through the identity recognition of varieties that are revealed by information. The value of the relational continuum and unity of matter, energy and information is continual creative destruction in the space if varieties.

The categories of values have different reflective understanding of the use and intrinsic values in ontological and epistemological spaces that must be distinguished. The act of distinguishing implies the existence of varieties defined by a distribution of characteristic-signal dispositions. Similarly, decision and choice imply the existence of varieties with inter-categorial and intra-categorial differences. In this epistemic structure, the ontological information is perfect in the sense that it is the same as knowledge which becomes input into the natural management of the command-control of natural transformations in terms of creation and destruction of categorial varieties. In this respect and over the ontological space, the intrinsic value and the use value of information are in relational continuum and unity. Such a relationality must be viewed in terms of ontological dynamics in the actual-potential polarity, where existing ontological varieties may be destroyed and new varieties may be created as substitute in the construction-destruction process. This is the essence of ontological creative destruction where matter and energy are in continual stock-flow equilibrium and information is in continual stock-flow disequilibrium. In this respect, the intrinsic value and the use value of ontological information are the same.

The epistemological information, however, is defective and this defectiveness may be enhanced by deceptiveness of cognitive agents with intentionalities relative to the management of command-control actions in the decision-choice systems for distribution of advantage that may come to affect the use value of information. The epistemological information is not equal to knowledge, and in fact, it substantially deviates from knowledge the quality and quantity of which depend on the distributions of the degrees of defectiveness formed by cognitive limitations and deception formed by intentionalities of decision-choice agents. Under the conditions of defectiveness and deceptiveness in the source-destination information process, the epistemological information cannot directly serve as input into the decision-choice process in the management of command-control actions for the dynamics of social actual-potential polarities. It becomes an input into the processing action within the constructionism-reductionism duality for the creation of social knowledge. The intrinsic value deviates from the use value where analysis may be done to find the conditions that are required for the equality. One thing that is clear, is that social knowledge just like information is always in continual stock-flow disequilibrium [R17.15, R17.18]. In respect of this, what does it mean in terms of information to claim that the universe is expanding? Is the actual universe expanding or is our knowledge about the universe expanding? Can one infer from the expansion of our knowledge that the actual universe is actually expanding?

5.2.1 The Use Value of Epistemological Information

The analytical concern about the value of epistemological information is on the use value as it relates to the dynamics of the social actual-potential polarity. The dynamics of social actual-potential polarity is seen in terms of revolution against the existing variety of social order and a contest of all potential social varieties under the general principle of internal action. The discussion on the relational concepts of duality, dualism and polarity where there is a distribution of relative negative-positive characteristics in establishing categorial varieties are offered in [R17.15, R17.16, R4.13]. The discussions here shall center on elements that are essential for the understanding of the present analytical system of the general theory of information. For any actual-potential polarity, the actual is a *singleton set* of a variety while the potential is an *infinite set* of varieties. The logic of social transformations is to intentionally and simultaneously potentialize the actual and actualize a potential as a substitute. The use value of any epistemological information must be seen in terms of possibilities of quality of knowledge production and decision-choice outcomes with and without the particular information. Here, the relational structure among information, knowledge and decision-choice actions must be kept in mind.

Information is an input into knowledge production given the paradigm of reasoning. The result of knowledge production is an input into decision-choice actions given the preference structure and some rationality. The results of decision-choice actions become inputs into information-updating. Epistemological information, therefore is an indirect input into social decision-choice actions through the knowledge process for continual creation, destruction and transformation of varieties and categorial varieties in the social space. This is unlike ontological information which is a direct input into the ontological decision-choice actions for continual creation and transformation of varieties and categorial varieties in the natural space. This point must be understood. In the ontological space, there is equality between information and knowledge, where information and knowledge are the works of the same hand and are used by the same hand for transformative actions where the source-destination information process is purely noiseless.

These conditions involving singularity of participant and information-knowledge equality do not necessary hold over the epistemological space where the input into the decision-choice system is indirect information through knowledge-construction, and not direct information because of the existence of information-knowledge inequalities. The claim of the importance of information to the decision-choice system holds only through the quality of the derived knowledge that may be obtained. Technically, decision-choice action is not a direct information process over the epistemological space. It is, however, a direct information process over the ontological space. Every decision-choice action over the epistemological space is a knowledge process given a decision-choice rationality, while knowledge is an informational process given the paradigm of thought. Information may be substituted for knowledge in the same paradigm of thought if there is a strong belief that the epistemic equality holds at a particular decision-choice time point where in this case there is assurance of noiseless transmission. The knowledge production process is such that the paradigm of thought is applied to the received signal disposition give its representation in order to construct the characteristic signal disposition that presents the identity of the variety. In other words, the knowledge-production process abstracts the epistemological characteristic disposition of a variety from the received ontological signal disposition by methodological constructionism. The epistemological characteristic disposition is then compared to the ontological characteristic disposition by methodological reductionism to see the degree of equality and abstract an *epistemic distance* to claim knowledge with possibilistic conditionality and probabilistic conditionality to deal with the problem of noisy source-destination transmission.

Given the same paradigm of knowledge production, the value of information must be viewed in terms of the relative possibilities of the decision-choice outcomes of an event with and without the information. The value of information, therefore, is seen in terms of differences in possibility measures of with and within transformations in the dynamics of social actual-potential polarity in the decision-choice actions. The analytical structure is similar to the framework of the cost-benefit analysis of one element in the choice set whose cost-benefit impact on the system is examined in terms of with the element and without the element. It is useful to keep in mind that at least in the social space, transformation involves the creation and destruction of varieties and categorial varieties in two steps. The first step of transformation is defined by understanding the set of *necessary conditions* that establishes *degrees of convertibility* the analysis of which defines conditions of *possibilistic uncertainties* and distribution of measures of possibility. The second step of transformation is defined by creating a set of *sufficient conditions* that establishes *degrees of likelihood of convertibility* given the distribution of the necessary conditions of the degrees of convertibility. The analysis of the degrees of likelihood of convertibility defines conditions of *probabilistic uncertainties* and distribution of measures of probability that will bring the actualization of a possible potential variety. The value of information, therefore, finds meaning, definition and measure under the cost-benefit rationality and Asantrofi-anoma principle in the decision-choice space [R5.13, R5.14].

A question arises as to whether the concept of the value of information is quantitative or qualitative or both in the sense that, it is defined within qualitative-quantitative duality and hence belongs to objective-subjective duality with relational continuum and unity. Pure qualitative concept at one extreme means that the value of information belongs to the value distribution of the subjective phenomena in relation to the conditions of preferences for either an individual, collective or both in the decision-choice space and that the analysis of the value of information belongs to methodological duality without an excluded middle. Pure quantitative concept at the other extreme means that the value of information is independent of subjective preferences of either the individual or collective in the decision-choice space and that the analysis of the value of information belongs to methodological dualism with an excluded middle. The combined qualitative-quantitative concept within the two extremes of the value of the epistemological information introduces a complexity in that one is dealing with the analysis in methodological duality with relational continuum and unity, which means that every qualitative value has a supporting quantitative value and every objective evaluation that can be extracted without preference structure has a supporting subjective evaluation that rests on preference ordering in the decision-choice space for transformations and creation of epistemological varieties.

5.2.1.1   The Source-Destination Asymmetry and Symmetry of the Value of Information

These different versions of the concept of the value of information is epistemological and not ontological. There is a multiplicity of the value of information over the epistemological space in which decision-choice agents operate. The multiplicity of the value of epistemological information creates complexity in the value-assessment problem between the source and destination in the transmission mechanism. The complexity is induced by intentionalities of sender-receiver modulus of a particular information type in the source-destination duality with relational continuum and unity in the transformational sense of important negation, where the source (sender) becomes the destination (receiver) and the destination (receiver) becomes the source (sender). The complexity shows itself in two inter-related forms as conditions of source-destination conflicts such as deception in war, diplomacy or antagonistic games, and also in conditions of source-destination cooperation such as intelligence sharing in war, negotiations, cooperative games and collaborations. In the case of conditions of conflict that appear in antagonistic relations, the information takes on a negative deceptive component that tends to increase the real cost of information to the receiver, and increase the benefit of the same information to the sender in decision-choice actions in the conflict space of cost-benefit duality in relational continuum and unity, where there is cost-benefit asymmetry of the values of the same information. The concepts of cost and benefit must be seen in terms of real variables of opportunity cost, where that which is destroyed is a loss of an old benefit and that which is created out of it is a gain of a new benefit. The loss of an old benefit is the cost of the gain of a new benefit in transformation. The cost and benefit, therefore reside in the same variety in terms of duality with relational continuum and unity.

In the case of conditions of cooperation, particularly in games of negotiations and war, the information assumes a positive deceptive component that increases the benefit of the same information to both the source (sender) and destination (receiver). This is positive information sharing that produces symmetry of value of the same information to the source and destination in decision-choice actions. The same information may be cost or benefit to the sender and receiver alike depending on intentionalities. Analytically, it is here and under conflict and cooperative zones that noises due to deceptive-information transmissions create their greatest effects through increasing or decreasing uncertainties, and affect the outcomes of decision-choice relative to the dynamics of social actual-potential polarity. It may be useful to keep in mind that every decision-choice action is about the creation-destruction process of varieties with real cost-benefit balances for continual updating of the information stock.

5.2.1.2 The Sender-Receiver Duality and Conditions for Measuring the Value of Information

In the sender-receiver duality, the value of information to the sender must be seen in terms of its effect on the knowledge production and decision-choice action of the receiver in relation to the sender's intentionality. Under sender-receiver cooperative conditions if the received information enhances the receiver's chances through improved knowledge for accomplishing an already decided goal then there is a *positive value* to both the sender and receiver. Under this condition, there is a reduction in both uncertainty and ignorance. Alternatively, if it reduces the receiver's chances through reduced knowledge for accomplishing an already decided goal then there is a *negative value* to both the sender and receiver relative to sender-receiver intentionalities. Under this condition, there is an increase in both uncertainty and ignorance. It may be noted that the value of information defined in terms of characteristic-signal disposition that projects varieties is symmetric to both the sender and receiver. Under conditions of sender-receiver rivalry, if the received information is enhanced through improved knowledge for accomplishing the receiver's already decided goal, then there is a positive value to the receiver and a negative value to the sender with positive intentionalities. Under this condition, there is a reduction in both uncertainty and ignorance to the receiver contrary to the sender's deceptive intentionality.

The values of epistemological information to the sender and receiver are reversed if the received information enhances the receiver's difficulty through reduced knowledge for accomplishing the receiver's predetermined goal. Under this condition, there is an increase in both uncertainty and ignorance to the receiver and improved benefit to the sender. The value of information is negative to the receiver and positive to the sender relative to the sender-receiver intentionalities. This type of source-destination information value system is an epistemological and not ontological phenomenon. The sender-receiver distribution of the value of information must be related to increase or reduction in uncertainties and ignorance relative to intentionalities and the existing defective information structure. The sender-receiver distribution of values may be related to social information policies of different nations where domestic and international social propaganda have the objective of manufacturing substantial disparities between characteristic disposition and signal disposition for the receiver in order for the sender to create advantage in the decision-choice actions and outcomes in the social space. In this respect, there is an asymmetry of the value of information in the source-destination structure due to the nature of sender-receiver intentionalities. The effectiveness of deceptive information structure in achieving the intentionalities depends on the receiver's social trust of the source and analytical power on the basis of a particular paradigm of thought. It is in this context that a critical examination of the relationship between information and games over the epistemological space become analytically contributing and practically useful.

5.2.1.3 Conditions for the Measurement of the Value of Information

There is an analytical problem which finds expression in the question of how does one measure the symmetric and asymmetric sender-receiver values of information. This is a value measurement problem of the sender-receiver utility in information configurations. The solution to this value measurement problem may be approached in two ways of possibility and probability frameworks. The possibility framework relates to possibilistic uncertainties and fuzzy variable, while the probabilistic framework relates to either stochastic or fuzzy uncertainties and hence to either random or fuzzy-random variables in the information-decision actions. It is also here that the concepts of possibility and probability spaces and their corresponding quality-quantity dispositions present some analytical difficulties. Some of these analytical difficulties have been discussed in [R4.10]. At the level of possibility, one is confronted with possibilistic uncertainties associated with the dynamics of actual-potential polarity, where the actual is to be either destroyed or transformed, and a potential is to be created as a substitute within the substitution-transformation process. Under the conditions of the dynamics of actual-potential polarity, the actual variety is to be destroyed or transformed and a potential variety is to be created as a substitute within the socio-natural transformation-substitution processes under the conditions of probability. Here, every probability is a potential and every potential is a possibility in degrees of actualization within any given environment as defined by a set of conditions of transformation of a particular phenomenon.

The degrees of transformability of the elements of the potential in relation to a particular actual are used to construct a possibilistic set and non-possibilistic set. The possibilistic set contains the relevant elements with different degrees of probabilistic uncertainties for the actualization by socio-natural action that is relevant to the intended decision-choice actual to be destroyed or transformed. The possibility space P may be decomposed into phenomenon-specific possibilistic sets, Π_ϕ where ϕ is a phenomenon that belongs to the infinite set of phenomena Φ with an index set $\mathbb{I}_\Phi^\infty$. The phenomenon-specific possibility set is a subset of the possibility space where $\Pi_\phi \subset P$ and $\bigcup_{i \in \mathbb{I}_\Phi^\infty} \Pi_{\phi_i} = P$. The possibility set defines the boundaries of the probabilistic set which is relevant to the actualization of a potential for a targeted phenomenon. In other words, there is a transformation movement from the possibility space with possibilistic uncertainties expressed by a membership distribution function to the probability space with probabilistic uncertainties. It may be made clear that the phenomenon-specific possibilistic set defines intra-categorial varieties with a possibility distribution, where such varieties are fixed by the distribution of characteristic-signal dispositions. The possibilistic set fixes the boundaries of applicable probabilistic reasoning by defining impossible and possible elements of actualization with probability distribution for decision-choice action. In this way the probability space is defined for all $\phi \in \Phi$. In assessing the relevant values of information, one may think of possibility as establishing the necessary conditions and probability as establishing the sufficient conditions, where the necessary conditions relate to necessity and the sufficient conditions relate to freedom in the decision-choice systems.

At the level of probability in the dynamics of actual-potential polarity, one is confronted with probabilistic uncertainty in either actualizing a potential element from the possibility space or potentializing the element in existence. The question, therefore, is whether the sender-receiver values of information in the source-destination process must be assessed in terms of reduction in either possibilistic or probabilistic uncertainties or both. One thing that is clear in this general theory of information that is being advanced is that the meaning of the concept of possibility and probability and their phenomena find expressions in the defined concept and phenomenon of information. Any distribution of the probability measures is supported by a distribution of possibility measures where each element in the probabilistic set belongs to it with a defined possibilistic degree of belonging, which together constitute the possibility distribution of a set of elements of a particular phenomenon. This set of elements then becomes an *event* with a distribution of probabilities for actualization. Any possibilistic set is a transformation set that defines the *necessary conditions* of uncertainty for transformation of a variety or creation of a new variety for the outcome of an event in terms of change.

The probability distribution of the probabilistic set defines the sufficient conditions of uncertainty for the event to be actualized. In this respect, the possibilistic uncertainties are enshrined in *necessity* while the probabilistic uncertainties are enshrined in *sufficiency* of transformation in the dynamics of actual-potential polarity. The creation of the required sufficiency is decision-choice dependent and hence generates *freedom*. It is in this context that one finds *freedom* and *necessity* working together in the dynamics of actual-potential polarity. It is useful to keep in mind that a necessity defines a set of conditions about a set of varieties of objects with defined characteristics and not others that must be acted on by socio-natural decision-choice activities to bring about an actualization of a particular phenomenon and potentialization of a particular actual variety and not others in the general dynamics of the actual-potential polarity. The set of the necessary conditions defines the necessity inherent in any actual variety the characteristic-signal disposition of which presents information on the type of categorial variety. The sufficiency defines a set of conditions about a set of varieties of socio-natural decision-choice actions with defined characteristics and not others that must be used for management of command and control of the objects of necessity to bring about an actualization of a particular phenomenon and potentialization of a particular actual and not others in the general dynamics of the actual-potential polarity. The set of the sufficient conditions defines the freedom of decision-choice agents in relation to any actual the characteristic-signal disposition of which presents information on the type of actualized event in a categorial variety. For extensive discussions on socio-natural actual-potential polarities and dualities see [R4.13, R17.15, R17.16, R17.18].

Some important question arises as to how the possibilistic set and probabilistic sets are analytically obtained or constructed and how they relate to the transformational dynamics of socio-natural actual-potential polarity. The search for the answers to these important questions on the organic information-knowledge decision-choice process led to the development of the monographs [R1.15, R4.10, R4.13, R17.16]. It is useful to keep in mind that the dynamics of actual-potential polarity pass through the possibility space from which the necessary conditions are obtained and then through the probability space from which the sufficient conditions are obtained in the substitution-transformation process. The process is such that there is the potential space U of universal existence, then there is the possibility space P in relation to the potential elements, next is the probability space B in relation to the possibilistic elements, and then the actual space A in relation to the probability space, where outcomes are realized as actual in the space of the actual from the potential through the possibility and probability spaces. The following relation holds, $A \subset B \subset P \subseteq U$ in terms of theories of acquaintance, informing, knowing, learning, teaching deciding and choosing. The necessary and sufficient conditions are the works of information in relation to the actual and the potential. The necessary conditions provide information on the varieties of objects to be manipulated and controlled to actualize a potential element. The sufficient conditions provide information on the decision-choice varieties of actions that must be used for the management of command and control to actualize the potential and potentialize an actual. The element from the potential is known to be actualized from the information on the content of the nature of its variety as established by its characteristic-signal disposition whose information goes to update the existing information stock.

5.2.1.4 Information Requirements and the Construction of Possibilistic-Probabilistic Sets

There are some discussion on the on the potential space U, the possibility space P, the probability space B and the space of the actual A as organic categorial varieties, each of which is defined by a collection of characteristic-signal dispositions that presents their information structure. The relational structure of continuum and unity among them has been explained. It has been stated that every element in the potential space is possible to be actualized and hence belongs to the possibility space with a distribution of degrees of convertibility or transformability that establishes the necessary conditions for conversions of varieties. A question then arises as to how the possibilistic set for a particular phenomenon is constructed from the potential space. The set of elements in the possibility space is infinite but the set of relevant elements that meet the necessary conditions of an actual-potential convertibility for any given phenomenon is finite. The general conditions of possibility of conversion of the varieties are used to construct degrees of possibility of actualization and ranking by the use of the fuzzy membership characteristic function. The use of the fuzzy membership characteristic function is justified by the presence of the high component of subjectivity in the subjective-objective duality of the phenomenon of possibility. It is always useful to keep a logical eye on the concepts of variety and characteristic-signal disposition on the basis of which order and ranking are objectively and subjectively established.

For each phenomenon, there is a set of possible elements with needed variety that satisfies the necessary conditions for actualizing it. Let this set of elements be Π_ϕ where ϕ is a generic element of the set of all phenomena Φ with infinite elements such that $\phi \in \Phi$ and $\Phi \subseteq P$ of the possibility space. Similarly, the set of all phenomena is also in the potential space such that $\Phi \subseteq U$ and $\Phi \subseteq U \supseteq P$. The set $\Pi_\phi \subseteq P$ is a finite possibilistic set for the phenomenon $\phi \in \Phi$ from which a selection is made for the actualization of the phenomenon of interest. All the elements in Π_ϕ meet the necessary conditions and are ranked by the fuzzy membership characteristic function. The possibilistic set Π_ϕ is equipped with the fuzzy membership characteristic function $\mu_{\Pi_\phi} \in (0,1)$ to create the necessary conditions for transformation. Thus,

$$\Pi_\phi = \left\{ \left(\pi, \mu_{\Pi_\phi}(\cdot) \right) | \pi \in \Pi_\phi, \mu_{\Pi_\phi}(\cdot) \in (0,1) \right\} \qquad (5.2.1.4.1)$$

The sufficient conditions for the actualization of the potential phenomenon reside in the probability space to which a fixed-level set of the relevant probabilistic set Π_ϕ belongs, and where the fixed-level is $\alpha \in (0,1)$ such that $\mu_{\Pi_\phi}(\cdot) \geq \alpha \in [\alpha, 1)$. By using fuzzy decomposition, the possibilistic set Π_ϕ, where $\phi \in \Phi$ may be decomposed into a crisp possibilistic set and a crisp non-possibilistic set where all elements belonging to the crisp possibilistic set have unit value of the possibility of convertibility.

The crisp non-possibility set is the defined fixed-level cut of the form $\mu_{\Pi_\phi} \in [0, \alpha)$ and every element in the crisp non-possibility set is considered as the impossible relative to the phenomenon $\phi \in \Phi$. These conditions allow one to define the crisp possibility set $\widehat{\Pi}_\phi$ and crisp non-possibility set $\widehat{\Pi}'_\phi$ as:

$$\left. \begin{aligned} \widehat{\Pi}_\phi &= \left\{ \left(\pi, \mu_{\Pi_\phi} \right) | \pi \in \Pi_\phi, \phi \in \Phi \text{ and } \mu_{\Pi_\phi}(\cdot) \geq \alpha \in (0,1) \text{ or } \mu_{\Pi_\phi}(\cdot) \in [\alpha, 1) \right\} \\ \widehat{\Pi}'_\phi &= \left\{ \left(\pi, \mu_{\Pi_\phi} \right) | \pi \in \Pi_\phi, \phi \in \Phi \text{ and } \mu_{\Pi_\phi} < \alpha \in (0,1) \text{ or } \mu_{\Pi_\phi}(\cdot) \in [0, \alpha) \right\} \end{aligned} \right\}$$
$$(5.2.1.4.2)$$

The elements in the crisp possibilistic set meet the *necessary conditions* for converting the potential $\phi \in \Phi$ to an actual $a \in A$. The crisp possibilistic set of the phenomenon $\phi \in \Phi$ becomes the *probabilistic set* $\mathbb{B}$ that defines the *sufficient conditions* with the probability distribution for the actualization of the phenomenon $\phi \in \Phi$ where $\mathbb{B} \subset B$ which may be written as:

$$\mathbb{B} = \left\{ (\pi, p) | \pi \in \widehat{\Pi}_\phi, \phi \in \Phi \text{ and } p = \varphi(\pi) \in (0,1] \right\} \subset B, \qquad (5.2.1.4.3)$$

where $\left(\widehat{\varPi}_\phi \cup \widehat{\varPi}'_\phi \right) = \varPi_\phi \subset P$ and $\phi \in \varPhi$. Every element in this discussion meets the conditions of variety and categorial varieties which are established by characteristic-signal disposition that informs the presence of the potential, possibility, probability and actual for knowing and learning. The sufficient conditions for the actualization of the relevant phenomenon is established by the probabilistic set where the elements in $\mathbb{B}$ are ordered in degrees of sufficiency by a probability distribution function $\varphi(\pi)$ in relation to the distribution of the characteristic-signal disposition (information) of the varieties in the set $\mathbb{B}$. For all elements $\pi \in \widehat{\varPi}'_\phi$, the probability is zero such that $\varphi(\pi) = 0$ if $\pi \in \widehat{\varPi}'_\phi \subset P$.

The phenomenon $\phi \in \varPhi$ is the event and the element $\pi \in \widehat{\varPi}_\phi$ is the possible outcome that can actualize the relevant variety when the socio-natural decision-choice actions bring about the sufficient condition relevant to a particular $\pi \in \widehat{\varPi}_\phi$. In constructing the probabilities or choosing the probability distribution for the outcomes in the probabilistic set, it is useful to keep in mind that the elements are distinguished in varieties, and their probabilities for actualization are established by their characteristic-signal dispositions. Let us also keep in mind that the set of characteristics for any object, broadly defined, is known over the ontological space through its signal disposition. In analytical works on informing, knowing, learning and decision-choice actions over the epistemological space, the signal dispositions are inexact but may be taken as exact by assumption. Under conditions of exactness of the signal disposition from the principles of acquaintance, there is an exact probability and hence the random variable is of the classical type the analysis of which requires the use of the classical paradigm of thought.

Under conditions of inexactness of the signal disposition from the principles of acquaintance, there is a fuzzy probability and hence with a corresponding fuzzy-random variable which is of the non-classical type. The analysis of the fuzzy-random variable requires the use of the fuzzy paradigm of thought. Both of which may be used in assessing the appropriate socio-natural decision-choice action or actions to create the sufficient conditions for the actualization of the potential phenomenon in the sense of creation of new varieties or the potentialization of the existing actual in the sense of destruction of existing varieties. It is useful to remember that the destruction of an existing variety creates a temporary vacuum that may be filled with an unintended variety as the new actual. This is the meaning of unintended consequences in socio-natural transformations. In this case, the socio-natural decision-choice actions were not compatible with the required sufficient conditions for the actualization of the desired potential. The unintended consequences are characteristics of decision-choice activities under the principle of intentionality in the epistemological space and not in the ontological space which constitutes the identity of universal existence.

5.2.1.5 The Possibilistic-Probabilistic Sets, Information and Actualization of a Potential Variety

The possibilistic set helps to define the necessary conditions of the actualization of phenomena from the potential space to the actual space, and from where each phenomenon meets the possibility conditions but it may not meet the necessary conditions for actualization. The necessary conditions are expressed through the distribution of characteristic-signal dispositions which indicate the varieties that must be acted upon, in both the potential space and the possibility space. At the level of epistemology, the set of the necessary conditions required for the transformation of any given phenomenon is the natural conditional elements that may be created by decision-choice agents. These necessary conditions are defined by socio-natural forces. It is here that necessity places limitations on freedom through sufficiency the understanding of the interactions of which requires information and knowledge. The probabilistic set helps to define the sufficient conditions of the actualization of phenomena from the potential space to the actual space, and from where each phenomenon meets the possibility conditions and the necessary conditions for actualization. The sufficient conditions are expressed through the distribution of characteristic-signal dispositions that affirm the varieties in both the potential space and the possibility space. At the level of epistemology, the set of the sufficient conditions for any given phenomenon is the creative work on information by the decision-choice agents, and hence is under the constrained management of command and control of the required actions for actualization. This is where freedom finds expression in choice of varieties and actualization of potential varieties as well as the destruction of existing varieties. Correspondingly, necessity finds expression in the limitation on freedom where both necessity and freedom find their respective expressions in information.

Given the necessary conditions as established by the possibilistic set and the sufficient conditions as established by the probabilistic set, a number of conceptual problems arise in the transformation process. These concepts revolve around the sufficiency conditions and the probability space. They include anticipation, uncertainty, expectation, ignorance, surprise and belief. These concepts claim their meanings and representations in the sufficiency conditions and freedom of command-control actions by decision-choice agents working with information given matter and energy. The meanings and contents of these concepts find their qualitative expressions from information on the basis of which their quantitative expressions can be constructed. The behaviors of these concepts amplify the restrictions on freedom of decision-choice actions to create the sufficiency conditions required for actualizing the needed potential. Interestingly, these concepts reflect the cognitive deficiency of decision-choice agents. The concepts of the potential, possibility and necessity in relation to information as defined by the characteristic-signal disposition relative to a variety, categorial varieties and possibilistic beliefs in the creative-destructive process in the actual-potential polarity have been discussed. The concepts of probability and sufficiency in relation to information as defined by the characteristic-signal disposition relative to a variety, categorial varieties and the probabilistic belief system in the creative-destructive process of varieties in the actual-potential polarity have also been discussed. The understanding and distinguishing of certain essential concepts in the quality-quantity space relative to events and outcomes require that the event must be placed in a proper space of either possibility or probability with a well-defined phenomenon $\phi \in \Phi$ so as to be aware of the types of uncertainty and risk that are definable in the epistemological space. Let us be reminded that there are three important universal elements of matter $\mathbb{M}$, energy $\mathbb{E}$ and information, $\mathbb{Z}$ existing in relational continuum and unity that define the universal space Ω where everything in $\mathbb{M}$ and $\mathbb{E}$ is potentially knowable in the epistemological space through $\mathbb{Z}$ that allow the establishment of varieties of matter and energy. The process of knowability is initialized by acquaintance.

5.2.1.6 A Simple Algebra of Possibilistic-Probabilistic Sets, Information and Actual-Potential Spaces

The process of information conceptualization involving socio-natural events of states, processes and transformations may be seen in terms of a simple algebra of set and category representations. First, the universal space Ω is made up of space of actual and potential space

U = Potential space with a generic element $u \in U$
P = Possibility space with a generic element $p \in P$
B = Probability space with a generic element $b \in B$
A = Space of the actual with a generic element $\pi^a \in A$
Φ = Space of phenomena with generic element $\phi \in \Phi$
Π_ϕ = A general possibilistic set for any phenomenon $\phi \in \Phi$ in degrees of belonging

$\widehat{\Pi}_\phi \subset \Pi_\phi$ = Crisp possibilistic set with increasing α-level cut relative to the actualization of $\phi \in \Phi \widehat{\Pi}'_\phi \subset \Pi_\phi \subset P$ = Crisp non-possibilistic set with degrees of possibility less than α relative to the actualization of $\phi \in \Phi$ where $\left(\widehat{\Pi}_\phi \cup \widehat{\Pi}'_\phi \right) =$
$\Pi_\phi \subset P$ with $\widehat{\Pi}_\phi \cap \widehat{\Pi}'_\phi = \emptyset$ and $\widehat{\Pi}_\phi \subset \widehat{B}$

$\widehat{B} = (B, \mu_B(\pi))$ = The fuzzy probability space equipped with degrees of inexactness that is a fuzzified crisp probability space.

There are a number of interrelated properties about this representation in relation to general theory of information and info-dynamics in the universal continuum and unity. First, the universal space Ω on which nature operates is made up of the space of the actual A and the potential space P such that $(U = P \cup A)$. This space is represented as in Fig. 5.1. It is the ontological space in which nature operates with the perfect information structure and no uncertainties on phenomenal elements of

Fig. 5.1 Categories of actual and potential spaces

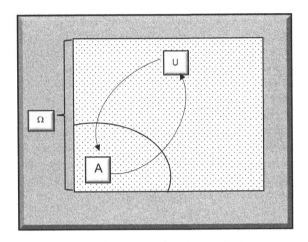

$\phi \in \Phi = \Omega$. In other words, every phenomenon is either actual or potential and every phenomenon $\phi \in \Phi$ sits in actual-potential polarity ready to be actualized or potentialized by an internal transformation process of relational continuum and unity of matter, energy and information.

The potential U is also the collection of all the phenomena in Φ which have not been actualized. The actual is the collection of all phenomena in Φ which have not been potentialized. Every element in either the potential or actual space is called a variety and every variety is associated with a phenomenon. The elements in the potential are all possible for actualization, while the elements in the actual space are also possible for destruction and potentialization and hence the possibility space P is the collection of all elements in the potential space with P = U as the analytical pivot point. This equality may be justified by keeping in mind that destruction is a potential and creation is also a potential in the dynamics of a system of actual-potential polarities. Socio-natural transformation of actual-potential polarities is a never-ending process under the conditions of the dynamics of creative destruction and substitution-transformation processes, where matter and energy are in stock-flow equilibrium, matter and energy cannot be created or destroyed. Information is in stock-flow disequilibrium, where every actualized potential is accompanied by a destroyed (potentialized) actual in the dynamics of actual-potential polarity. The dynamics of actual-potential polarities work through the interplay of a series of residing dualities, where the recognition of any transformation is through an acquaintance with the corresponding signal disposition that reveals the underlying characteristic-signal disposition and the differences between the old and new varieties of the same object. The comparative analysis is about the dated varieties of the same object under transformation. In other words, statistical analysis or big data analytics is analysis of quantitative varieties as seen through the distribution of signal dispositions in the epistemological space.

The information as defined by characteristic-signal disposition is complete and non-fuzzy in the ontological space in the sense that there is non-defectiveness and non-deceptiveness in the natural decision-choice process for transformations. In this respect, nature transforms itself with perfect information where the necessary and sufficient conditions of transformation are generated from the source and destination of the same entity in an inseparable existence. This condition of perfect information *eliminates the logical existence* of the probability space in the natural decision-choice management of command-control actions in the transformation of varieties since there are no uncertainties but a direct link between the actual and the potential in the dynamics of the system of actual-potential polarities. There are no cognitive limitations that can create uncertainties and risks. It is useful to keep in mind that every phenomenon in the actual space resides in the possibility of transformation into a potential through the destruction of an existing variety, and that every phenomenon in the potential also resides in the possibility of transformation into an actual through the creation of a new variety. The information on the destroyed variety remains and the information on the new variety is added to the stock of information.

The understanding of this natural process of the dynamics of actual-potential polarity by cognitive agents is through info-dynamics, and the use of this understanding by them acting as decision-choice agents takes place over the epistemological space which requires us to account for cognitive limitations that appear through the manufacturing and use of defective and deceptive information structures on the part of decision-choice agents. In other words, the space of the phenomena, the potential space, the possibility space, and the space of the actual cannot be connected smoothly as in the case of natural decision-choice actions in the dynamics of any actual-potential polarity where the actual and potential varieties are directly interdependent through information sharing. The defective and deceptive information structures show themselves in the probability spaces relative to any given phenomenon and the intentionalities of the decision-choice agents in actualizing a phenomenon as an outcome of a decision-choice action. The probability must be connected to the possibility which connects to all the relevant spaces. This connection must be constructed in the epistemological space as a logical tool to take account of the cognitive limitations. To account for the interconnectedness, it is observed that the possibility and probability spaces are introduced for info-dynamics of the stock-flow disequilibrium in order to connect the potential space P to the space of the actual A over the epistemological space. Since the possibility space is a collection of potential phenomena it is useful to partition it into respective possibilistic sets of the form, Π_ϕ where $\Pi_\phi \subset P$ and $\bigcup_{\phi \in \Phi} \Pi_\phi = P$ which is constructed in reference to the potential.

In natural transformational dynamics over the ontological space, the sufficient and necessary conditions are simultaneously handled in the possibility space since the structure of the info-dynamics is the supreme architect of nature from within and the transformation is from within in terms of the internal ability to create and destroy. This is not the case over the epistemological space where cognitive agents must be informed, must know and learn with limitations to affect a change. Cognitive agents have very little control, and in fact, have control over themselves with small freedom which is substantially constrained by necessity in terms of their actual and potential existence. Cognitive agents are at the mercy of natural forces in the creation-destruction process that provides the necessary conditions which limit the sufficient conditions. They have to know the *necessary conditions* and then manufacture the *sufficient conditions* through decision-choice action [R4.13, R17.15, R17.16]. These necessary and sufficient conditions reside in polarities and dualities under the general principle of opposites to generate energy, power and force of transformation from within. It is this principle of opposites that the game-theoretic approach finds usefulness in the analysis of socio-natural transformation.

To be informed of the necessary conditions and manufacture the sufficient conditions of transformation, it is observed that the necessary conditions reside in the possibility space which restricts freedom and the sufficiency conditions which reside in the capacity of cognitive agents in terms of restricted freedom to craft transformative actions to destroy an actual variety a $\in$ A and actualize a potential phenomenon $\phi \in \Phi$. Corresponding to each phenomenon ϕ, is a possibilistic set Π_ϕ, and hence there are as many possibilistic sets as there are phenomena. It is also observed that the elements in each possibilistic set are ranked in degrees of possibility in actualizing any phenomenon $\phi \in \Phi$. The possibilistic set Π_ϕ may then be decomposed into a crisp possibility set $\widehat{\Pi}_\phi$ and a crisp non-possibility set $\widehat{\Pi}'_\phi$ such that $\widehat{\Pi}_\phi \cap \widehat{\Pi}'_\phi = \emptyset$ and $\Pi_\phi = \left(\widehat{\Pi}_\phi \cup \widehat{\Pi}_\phi \right)$ with $\Pi_\phi \cap B = \widehat{\Pi}_\phi$, $(B \cap P) = \widehat{\Pi}_\phi$ and $\left(B = \bigcup_{\phi \in \Phi} \widehat{\Pi}_\phi \right) \subset \left(P = \bigcup_{\phi \in \Phi} \Pi_\phi \right)$. In other words, the possibility space always contains the probability space in the info-dynamics over the epistemological space. The structure may be imposed on the ontological space as represented in Fig. 5.2. This set structure provides an important analytical way of viewing some important defining concepts in information-decision-interactive processes to which we now turn our attention.

All *uncertainties* whether possibilistic, probabilistic or fuzzy are information-based concepts as well as epistemological and not ontological. The concepts of possibility and probability are defined in the quality-quantity duality in the sense that they contain qualitative-quantitative dispositions that define their specific varieties. The possibilistic uncertainty is due to information vagueness and hence finds expressions in *fuzzy variable* and *fuzzy event* and measured in fuzzy *space*, where in this case, the event $\pi \in \Pi_\phi \subset P$. The probabilistic uncertainty is due to information quantity limitation and hence finds expression in *random variables* and *random events* and measured in the *probability space* where in this case the event $\pi \in \widehat{\Pi}_\phi \subset B$. The simultaneous existence of possibilistic and probabilistic uncertainties finds expression in either *fuzzy-random variables* or *random-fuzzy variables*. The fuzzy-random or random-fuzzy variables are then measured in the *fuzzy-stochastic space* in some form of fuzzy complexification of the probability space equipped with a fuzzy characteristic membership function to express the distribution of the qualitative degrees of vagueness or inexactness, where in this case the event $\pi \in \widehat{\Pi}_\phi \subset \widehat{B} = (B, \mu_B(\pi))$ [Kaufman fuzzy arithmetic] [R4.13]. *Ignorance*

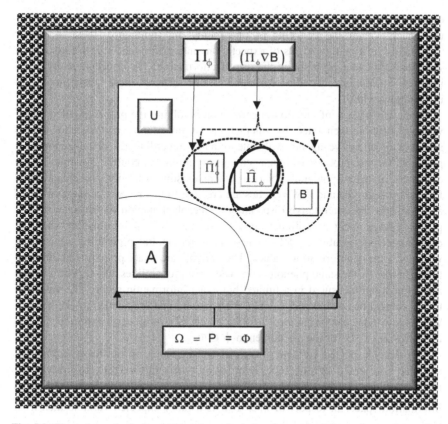

Fig. 5.2 The epistemological partition of ontological universe for information understanding toward knowledge construction and decision-choice actions

on the part of cognitive and decision-choice agents is a knowledge-based concept. It is also epistemological and not ontological. Knowledge is a derivative of information and hence uncertainty is equal ignorance to the extent to which information equals knowledge. Ignorance is an indirect concept from the lack of processed information by a justified paradigm of thought about a particular phenomenon $\phi \in \Phi$.

5.2.1.7   Information, Expectation, Anticipation and Surprise in Decision-Choice Systems

The general definition of the concept of information and the nature of the phenomenon of information are fixed in terms of varieties and categorial varieties as presented by the characteristic-signal dispositions and their distributions in time and over time. From the general information definition the concepts of uncertainty and ignorance are defined and explicated in relation to cognitive activities over the epistemological space. Uncertainty and ignorance are not found over the ontological space. Given the concepts of uncertainty and ignorance, there are other importantly relational concepts of expectation, anticipation and surprise associated with decision-choice systems in the use of information. The concepts require definitional restrictions in the epistemic frame where these restrictions are information-defined.

Expectation is an information phenomenon involving the probable outcomes of varieties that are generated by interactions of uncertainty and decision-choice activities of cognitive agents with intentionality regarding the actualization of a potential in the process of management of command-control systems in the dynamics of actual-potential polarity, where existing varieties are either transformed or destroyed and new varieties are created. In this respect, a question arises as to what is the quantitative disposition of the phenomenon of expectation. The process brings about a continual updating of the stock of information where an exact expected variable is defined and measured in the probability space as a weighted sum of information values. The weights are exact prior probabilities on the outcomes of the same phenomenon. The prior probabilities and expected values are defined and measured in a limited but exact information space. The values of these probabilities and expectations provide conditional boundaries on the basis of which knowledge can be claimed. In other words, if π^e is an exact expected variable then $\pi^e \in \Pi_\phi \subset \mathrm{B}$ for the phenomenon $\phi \in \Phi$.

The fuzzy expected variable is defined and measured in the fuzzy probability space as a weighted sum of fuzzy information where the weights are fuzzy probability measures. In other words, it is defined and measured in a vague and limited information space. In respect of this, if $\widehat{\pi}^e$ is a fuzzy expected variable then $\widehat{\pi}^e \in \widehat{\Pi}_\phi \subset \widehat{B} = (B, \mu_B(\pi))$ for the phenomenon $\phi \in \Phi$. It is useful to note that the vagueness and limitedness are always in reference to the signal disposition component of information as defined by the characteristic-signal disposition since the defective and deceptive information structures are the works of acquaintance and interpretation of the signals as they relate to the characteristics of the varieties. It may be kept in mind that acquaintance may take many forms such as different kinds of observation and experimentation with instruments which may enhance the observations by cognitive agents. The situation that $\widehat{\pi}^e \in \widehat{\Pi}_\phi$ simply means that the fuzzy-expected variable not only meets the necessary and sufficient conditions but it accounts for either subjectivity, qualitative disposition, inexactness or all of them in the algorithmic process in cognition.

Anticipation is an information-based phenomenon in decision-choice conditions involving fuzzy present-future time connectivity [R17.18]. Any anticipatory variable is defined by fuzzy-random conditions in a fuzzy probability space with fuzzy-probability distribution where there are inexact probabilities due to subjectivity, qualitative disposition or elements of inexactness. In other words, an anticipatory variable $\pi^a \in \widehat{\Pi}_\phi \subset \widehat{B}$ is defined with fuzzy probability specified as $\varphi\left(\pi^a, \mu_{\Pi_\phi}(\pi^a)\right)$. The anticipation is thus measured as a weighted expectation where the weights are defined by a fuzzy membership characteristic function that expresses the degrees of belonging to the anticipatory space which is a sub-space of the expectation space. The nature of the expectation space is such that there are some events in the space of expectation that are anticipatory and others that are not. These anticipatory events are defined in the fuzzy-stochastic-time space. By defining anticipatory events as represented by a fuzzy-random variable, one seeks to combine qualitative-quantitative dispositions in the subjective-objective conditions of the phenomenon of information and its uses in conceptualizing the complexity of present-future decision-choice outcomes as seen from the past in one's preference in the understanding of exact or fuzzy time connectivity. There is a close relation between anticipation and fuzzy expectation in decision-choice processes.

Surprise is also a knowledge-based phenomenon conditional on given information defined in the space of the actual relative to the impossibility space. It is the result of a knowledge deficiency due to either a limitation in the information processing capacity of decision-choice agents or due to both quantitative and qualitative limitation of information. In other words, a surprise is said to have happened if the event is actualized from the possibility space without satisfying the necessary and sufficient conditions of actualization. A *surprise variable* π^s is a possibilistic variable that does not belong to $\widehat{\Pi}_\phi \subset \widehat{B}$ but belongs to the space of the actual A. In other words $\pi^s \notin \widehat{\Pi}_\phi \subset \widehat{B}$ but $\pi^s \in A$. The measure of the concept of surprise is, therefore, through the possibility distribution. An actualized event is a surprise if its presence has no connectivity with the necessary and sufficient conditions for the actualization of a given phenomenon $\phi \in \Phi$. The surprise space contains variables the phenomena of which belong to the symmetric difference of the impossible set and possibilistic set as cognitively conceived through the best available methods of knowledge production. The surprise variable belongs to the impossible space relative to the phenomenon $\phi \in \Phi$. The conditions of surprise may easily be placed in the non-stochastic fuzzy space since the knowledge deficiency may be due to both quantitative and qualitative differences as revealed by the signal disposition of the phenomenon $\phi \in \Phi$. It is important to reflect on the notion that expectation and anticipation formations are done in the possibility space and have present-future time connectivity relative to decisions and realizations over the transformation space of varieties. Surprise is present-defined with realization in the space of actual and has past-present time connectivity in the transformation space of varieties. Surprise may be related to the prior information ignorance as conceived in the probability space.

5.3 Past-Present-Future Information and Decision-Choice Actions (The *Sankofa-Anoma* Problem)

The last section was devoted to discussions on the meanings of uncertainty, ignorance, expectation, anticipation and surprise as they relate to information. The meanings of these concepts and their quantifications find expressions relative to the meaning and concept of information and its phenomenon. Nothing else can help to define them. It was also pointed out that the role of information is found in *knowledge and decision*. These discussions apply to static and dynamic conditions as well as conditions of states and processes in socio-natural transformations. It was made clear that decision-choice actions are about changes and transformations of varieties of states and processes that relate to the dynamics of actual-potential polarities in socio-natural spaces where the varieties are defined by their characteristic-signal dispositions which present the information content and identities of the varieties. It was also argued that the basic nature of information is that it is indestructible and always being created in such a way that it is always in *stock-flow disequilibrium*.

The stock-flow disequilibrium of information is an indispensable and essential property of info-dynamics and info-processes. It is this nature of continual stock-flow disequilibrium that distinguishes information from matter and energy which are always in stock-flow equilibria. It is also this property that distinguishes info-dynamics from thermodynamics, electrodynamics, chemo-dynamics and other forms of material-based dynamics. It was also explained that information is not knowledge; it is simply an input into knowledge production over the epistemological space. The stock-flow info-dynamics must be related to the inputs of knowledge production while the results of the knowledge production become the inputs into decision-choice activities in managing ignorance, expectations, anticipations, surprises, and unintended consequences in relation to controllability and convertibility of varieties and categorial varieties in the general dynamics of the system of actual-potential polarities. The information related to the knowledge-decision processes is defined in time continuum in past-present-future relational unity, where all the current decisions involve the integrated processes of past-present duality and present-future duality with decision and transformation connectivity.

The present-past connectivity is backward looking on the basis of past history while the present-future connectivity is forward looking on the basis of future history. These two dualities generate information concepts of discounting forecasting, prediction and prescription relevant to all decision-choice systems involving management of control-command actions in continual information creation as seen in transformations of varieties and categorial varieties. The process of uniting the present-past and present-future dualities into the *time trinity* is called the *Sankofa-Anoma problem* in the African philosophical traditions in the Akan linguistics. It is important to reflect on the notion that expectation, anticipation and these concepts must have information definitional restrictions to which we now focus our attention.

5.3.1 On the Information Concept of Discounting

Discounting is a process of using a set of methods and techniques for examining the relevance of the future flow of information on the current stock of information on possible decision-choice actions that will create future varieties. It is a weighted composite sum of the future characteristic-signal dispositions of varieties which are brought to the present as input into current decision-choice actions relative to a particular phenomenon the quantitative disposition of which needs a change while holding the qualitative disposition constant. The weights are distributed between zero and one where each assigned weight decreases as one travels into distant future on the characteristic-signal disposition. In this way, the awareness of future possible varieties are connected to the present through the information-decision-interactive processes regarding a particular phenomenon. Discounting, therefore, connects the future to the present and bring them to the comparative past to revise the present information stock in terms of the quantitative disposition of a variety of a given phenomenon. The analysis is an *intra-categorial process* where the variety as defined by qualitative disposition is held constant and the past-future-present internal quantitative disposition is examined. The importance of the future infor-mation flow at each time point is examined relative to the current information on a variety. Since information is always in a stock-flow disequilibrium at each time point, some questions arise as to how far should one view the historic forward and what is the distribution of the degrees of importance that must be attached to the forward distribution of time-point characteristic-signal dispositions in the quanti-tative domain, holding quality constant.

The distribution of the degrees of importance presents itself as a distribution of time-point weights that will reveal the importance of each forward time-point quantitative characteristic-signal disposition on the decision-choice action to create future flow for the variety of the same phenomenon. In discounting, the distribution of the future time-point weights may be allowed to be revealed by the methods and techniques used to connect the distribution of the present-future characteristic-signal dispositions. The choice of the inter-temporal forward weights as the historic future is viewed will depend on current time-point preferences relative to the distant future. The choice of the distribution of the inter-temporal forward weights will also depend on whether the future varieties are collectively or individually owned. In other words there are individual and collective discounting factors which are not necessarily the same. There are many formulae that capture different distributions of weights in relation to the outcomes of future varieties the characteristic-signal dispositions of which will enter into the current decision-choice actions. Discounting is simply an information search on future varieties in terms of their relevance into current decision-choice from the futures varieties that will optimize greater socio-natural good. The optimality is seen through the distribution of the present-future signal dispositions and mapped onto the future characteristic dis-position to reveal the quantitative outcome of a preferred variety.

A discounted variable, π^d is such that $\pi^d \in \Pi_\phi \subset P$ and also $\pi^d \in \widehat{\Pi}_\phi \subset B$ for a given phenomenon $\phi \in \Phi$ and a corresponding object $\omega \in \Omega$ which simply implies that it is assumed to meet the necessary and sufficient conditions of actualization of the phenomenon and its object variety. It must be clear that discounting does not have any theory of either explanation or prescription and hence it is not prediction. The theory of discounting is about algorithms of connecting the quantitative future to the quantitative present and to the quantitative future in terms of information flows (info-flows). The discounting process may be combined with forecasted values and/or predicted values to produce discounted values as inputs into a current decision-choice action to create a future variety with a corresponding characteristic-signal disposition.

In other words, the inputs in the algorithms are the relevant elements of information in A for the phenomenon $\phi \in \Phi$. This may be called intra-categorial forecasting since the quality is held constant and the same qualitative variety is under focus. It will become clear that there are similarities and differences between discounting and forecasting. Each of these methods is to bring information as input into current decision-choice actions for the creation of future varieties over the epistemological space. The discounting is to bring the future information into the present decision-choice action to create new future while forecasting, as it will be shown, is to bring the reflections of the past information into present decision-choice action to create a future as seen in varieties and categorial varieties. Interestingly, prediction is to take the distribution of the past-present characteristic-signal dispositions and suggest what future varieties may be confronted. Discounting does not involve a change in variety in terms of qualitative disposition and hence one cannot speak of comparative analysis of $(\phi_i \in \Phi, \omega_i \in \Omega)$ and $(\phi_j \in \Phi, \omega_j \in \Omega)$. One does not have a logical way to speak of inter-categorial discounting which simply implies discounting a distribution of inter-temporal qualitative varieties until a common unit is created to allow aggregation. The future time-point weights connected to the assumed future decision-choice outcomes reflect the possible varieties and the prior information relative to the outcomes that may be realized in the future time point values as abstracted from the possibility space and projected into the space of the actual. The future time-point weights and prior probabilities, therefore, have logical and computational connectivity in decisions and social transformation under command-control systems.

5.3.2 On the Information Concept of Forecasting

Forecasting is a process of using a set of methods and techniques for examining the relevance of the stock of past and present information and past-present varieties on a possible flow of future information as projected by future varieties. It is a weighted composite sum of the past, present and forecasted flow of information, which may then be used as current information input into current decision-choice actions relative to a particular phenomenon the quantitative disposition of which needs a change while holding the qualitative disposition constant. Forecasting, therefore, connects the past to the future and brings them to the present to revise the present information stock in terms of the quantitative disposition of a variety of a given phenomenon. The analysis is an *intra-categorial process* where the variety as defined by

qualitative disposition is held constant and the past-future-present internal quantitative disposition is examined. The importance of the past stock of information at
each time point is examined relative to the future information on a variety.

Since information is always in a stock-flow disequilibrium at each time point,
some questions arises as to how far back should one go and what is the distribution
of the degrees of importance that must be attached to each past information time
point in the quantitative domain holding quality constant. The distribution of the
degrees of importance presents itself as a distribution of time-point weights that will
reveal the importance of each time-point stock of information on the future flow for
the variety of the same phenomenon. In forecasting, the distribution of the
time-point weights may be allowed to be revealed by methods and techniques used
to connect the past-present-future information. The weights, as obtained, are the
distributed lags which are nothing more than the distribution of the degrees of
importance of the effects of a past time-point stock of information on the quantitative variety of a given phenomenon, where the processes create information flows
to amplify the stock of the future info-stock. In quantitative disposition of varieties,
different lag-models have been constructed to find the appropriate and best weight
distribution of the time-lags for different varieties as one travels over time points of
the distant past. Forecasting is simply an information search on future varieties as
seen through the distribution of the past-present signal dispositions and mapped
onto the future characteristic disposition to reveal the quantitative outcome of a
variety. Forecasting is not directly related to the test of collaboration, verification
and falsification of either explanatory theory or prescriptive theory as mechanisms
of knowledge production.

A forecasted variable is such that $\pi^f \in \hat{\Pi}_\phi \subset B$ and also $\pi^f \in A$ for a given
phenomenon $\phi \in \Phi$ which simply implies that it is assumed to meet the necessary
and sufficient conditions of the actualization of the phenomenon. It must be clear
that forecasting does not have any theory of either explanation or prediction and
hence it is not prediction. The theory of forecasting is about algorithms of connecting the quantitative past to the quantitative present and to the quantitative future
in terms of information flows. In other words, the information inputs in the algorithms are on the relevant elements of the actual in A for the phenomenon $\phi \in \Phi$.
This may be called intra-categorial forecasting since the quality is held constant and
the same variety is under focus. When the forecasting involves a change in variety
in terms of qualitative disposition from $\phi_1 \in \Phi$ to $\phi_2 \in \Phi$, then an inter-categorial
forecasting is implied. The input conditions of inter-categorial forecasting may vary
from forecasting to another on the same phenomenon but abstracted from A.

5.3.3 On the Information Concept of Prediction

Prediction is a process of using a mechanism of the explanatory or prescriptive past
and present stock of information to examine a possible flow of future information as

may be used to describe the identities of outcomes of future varieties. It is a process that connects the current variety to the information on the future variety in terms of actualization of different varieties or repetition of the same variety. The prediction of an actualization of different varieties from the current variety is an *inter-categorial prediction* in the sense that the variety $\phi_1 \in \Phi$ with information $\mathbb{Z}_{\phi_1}$ will change into the future variety $\phi_2 \in \Phi$ with information $\mathbb{Z}_{\phi_2}$, while the prediction of repetition of the same variety is an *intra-categorial prediction* in the sense that the current variety $\phi \in \Phi$ with information $\mathbb{Z}_\phi$ will show itself as the future variety of $\phi \in \Phi$ with the same information $\mathbb{Z}_\phi$ under the same conditions. The intra-categorial prediction is explanatory-theory based where the contents of the quantitative disposition of the set of propositions implied by the explanatory theory constitute the necessary and sufficient conditions for projecting the future occurrence of the same phenomenon (variety) in a category of interest. The *explanatory mechanism* of the *explanatory theory* involves past and present information. Unlike forecasting, it is not a weighted composite sum of the past, present and future information flow.

The characteristic-signal disposition of the explanatory-based predicted variable as future information may be included in the stock of current information that may be taken as a current input into current decision-choice actions relative to a particular phenomenon $\phi \in \Phi$ the quantitative dispositions of which need changes while holding the qualitative disposition constant. In other words, if $\phi_1 \in \Phi$ is the current phenomenon with current information $\mathbb{Z}_{\phi_1}$ and $\phi_2 \in \Phi$ is the predicted phenomenon with predicted information $\mathbb{Z}_{\phi_2}$ then ϕ_1 must be the same phenomenon as ϕ_2 with $\phi_1 = \phi_2$ and $\mathbb{Z}_{\phi_1} = \mathbb{Z}_{\phi_2}$. In this respect, intra-categorial predictions may be interpreted as backward-written explanations in the methodological duality of constructionism and reductionism, where constructionism applies to the development of the explanatory theory, and reductionism applies to the predictive process to provide a corroborative power of the theory. An example of intra-categorial prediction may be explained by considering the category of high rates of economic growth H with a generic element $h \in H$ and the category of low rates of economic growth L with a generic element $l \in L$ within the category of less-developed countries U with a generis element $u \in U$. The intra-categorial prediction is information $\mathbb{Z}_h \subset \mathbb{Z}$ that shows how an element $l \in L$ is converted to $h \in H$ through intra-categorial conversion that may be represented as $l \in L \xrightarrow{1} H$.

The *inter-categorial prediction* is prescriptive-theory based, where the contents of the quantitative-qualitative dispositions of the set of propositions implied by the prescriptive theory constitute the necessary and sufficient conditions for projecting the future occurrence of a different phenomenon (variety) in a different category of interest. The prescriptive mechanism of the prescriptive theory involves an *axiomatic information stock*. The characteristic-signal disposition of the predicted variety as a future information flow may be included in the stock of current information that may be taken as a current input into current decision-choice actions relative to a particular phenomenon $\phi_2 \in \Phi$ with information $\mathbb{Z}_{\phi_2}$ the quantitative-qualitative dispositions of which need simultaneous changes. In this

respect, inter-categorial predictions may be interpreted as forward-written explanations of future variety in the methodological duality of constructionism and
reductionism, where constructionism applies to the development of the prescriptive
theory, and reductionism applies to the explanatory process to provide a prescriptive power to affirm the validity of the theory. In other words, if $\phi_1 \in \Phi$ with
$\mathbb{Z}_{\phi_1} \subset \mathbb{Z}$ is the current phenomenon and $\phi_2 \in \Phi$ with $\mathbb{Z}_{\phi_2} \subset \mathbb{Z}$ is the predicted
phenomenon, then ϕ_1 must not be the same phenomenon as ϕ_2 with $\phi_1 \neq \phi_2$ and
$\mathbb{Z}_{\phi_1} \neq \mathbb{Z}_{\phi_2}$. An example of inter-categorial prediction may be explained by considering the category of developed economies D with a generic element $d \in D$ and
the category of less-developed countries U with a generic element $u \in U$. The
inter-categorial prediction is information $\mathbb{Z}_u \subset \mathbb{Z}$ that shows how an element $u \in U$
is converted to $d \in D$ through inter-categorial conversion that may be represented
as $u \in U \xrightarrow{u} D$.

Both *inter-categorial* and intra-categorial predictions take place within the
dynamics of the system of actual-potential polarities. They are formal in their
analytical approach. There is also an informal predictive process about the occur-
rences of future varieties. Informal prediction is a process of trying to connect the
future flow of information to the present stock of information through more or less
guesswork on the basis of available time-points of past information stocks or on the
basis of some implied knowledge. An informally predicted variable π^{ip} for a
phenomenon $\phi \in \Phi$ is such that $\pi^{ip} \in (\Pi_\phi \nabla B) \subset P$ and $\pi^{ip} \in A$ which simply
means that it may not satisfy both the necessary and sufficient conditions for
actualization. However, the prediction is possible and probable under certain belief
conditions that must be examined. It is always useful to keep in mind that the space
of the actual presents current stock of known and unknown information and hence
any variable belongs to it whether predicted or forecasted such as $\pi^{ip} \in A$ with its
corresponding information is taken to be a flow addition to the stock of current
information on the basis of which it may serve as an input into the decision-choice
mechanism. The space of the actual is the collection of actual varieties as defined by
the distribution of characteristic-signal dispositions. The known portion of the space
of the actual at any time point is the stock of knowledge of known past-present
varieties, while the unknown portion of the space of the actual is pure *cognitive
ignorance* of past-present varieties as viewed through the complete distribution of
characteristic-signal dispositions over the quality-quantity domain. It is useful to be
mindful of the idea that the space of the actual A as seen over the epistemological
space is made up of the known actual A^k and the unknown actual A^u at any time
such that $A^k \cup A^u = A$ and $A^k \ll A^u \subset A$.

It may be analytically helpful to examine the similarities and differences between
the explanatory-theory-based predicted variable π^{ep} and the prescriptive-theory-
based predicted variable π^{pp} and place them in the proper space and subspace of the
information system that allows the establishment of varieties of linguistic, logical
and mathematical spaces. One thing of similarity is that both of them belong to the

space of the actual as viewed as information input and hence $(\pi^{\varepsilon p}, \pi^{pp} \in A)$, however $\pi^{\varepsilon p} \in \widehat{\Pi}_{\phi} \subset B$ for $\phi_1, \phi_2 \in \Phi$ with $\phi_1 = \phi_2$ as the same variety of phenomenon with the same characteristic-signal disposition.

5.3.4 On the Information Concept of Prescription

In the previous section, the notion of prescriptive theory was introduced to allow the introduction of the idea of axiomatic based prescription with an axiomatic information structure which was related to inter-categorial transformations in the quality-quantity space. The question then is how does one define and explicate the concept of prescription for use in the scientific analysis of transformation decisions. Prescription is a process is a process of designing a decision-choice path with rules to guide actions of decision agents in order to set a potential against an actual in relation to a phenomenon $\phi \in \Phi$ and a corresponding object $\omega \in \Omega$ by destroying the existing actual variety $(v \in A)$ with a characteristic-signal disposition of the form $\mathbb{Z}_{(v \in A)} = (\mathbb{X} \otimes \mathbb{S})_{(v \in A)}$ and bringing into existence a new actual variety $(\upsilon \in U)$ with a characteristic-signal disposition of the form $\mathbb{Z}_{(\upsilon \in U)} = (\mathbb{X} \otimes \mathbb{S})_{(\upsilon \in U)}$ from the potential space through the principle of the transformation-substitution process where $(\upsilon \in U)$ is substituted for the old actual $(v \in A)$. Prescriptions are a strategic set of rules and actions to be used to bring about the disappearance of an existing variety as defined by its characteristic-signal disposition in the space of the actual, and the emergence of a new variety as defined by its characteristic disposition in the potential space where desirable and undesirable varieties are cast in the actual-potential polarity. In this process, the set of the prescriptive rules and actions finds expressions in the duality of negative-positive characteristic sets on the basis of which characteristic-signal dispositions inform cognitive agents. The characteristic-signal dispositions present information about varieties in actual and potential space in time and over time, and the possibility and probability spaces help cognitive agents to link the actual and potential spaces for informing, knowing, learning and creating prescriptive decision-choice rules and actions for the management of command and control processes on the path of transformation-substitution activities for the dynamics of the social actual-potential polarities in destroying existing varieties in the actual space and creating new varieties from the potential space.

In the conduct of prescription, there are always two distributions of varieties that characterize the space of actual-potential polarity. They are the distribution of actual varieties $\mathbb{D}_A = \{a_i | a \in A \text{ and } i \in \mathbb{I}_A^{\infty}\}$ and the distribution of potential varieties of the form $\mathbb{D}_U = \{u_i | u \in U \text{ and } i \in \mathbb{I}_U^{\infty}\}$ that present conflicts for then continual game of transformation-substitution process that is played within the interactions between the negative-positive duality and the quality-quantity duality to create sufficient conditions for potentializing the existing actual variety and actualizing a potential variety. It is axiomatically useful to assume that $\mathbb{I}_A^\infty \subset \mathbb{I}_U^\infty$ and hence $\#\mathbb{D}_A < \#\mathbb{D}_U$. For the purpose of prescription, the set of actual varieties must be partitioned into the set of known actual varieties $\mathbb{D}_A^K$ and a set of unknown actual varieties $\mathbb{D}_A^U$ such that $\mathbb{D}_A = \left(\mathbb{D}_A^K \cup \mathbb{D}_A^U\right)$. The set of the known past-present actual varieties constitutes the stock of knowledge while the set of unknown actual varieties constitutes the state of cognitive ignorance. The axiomatic condition underlining the dynamics of the actual-potential polarity with the substitution-transformation supporting principle is that every actual variety is not only a candidate of being potentialized but is under a continual process of being transformed. Similarly, every potential variety is not only a candidate of being actualized but is in a continual process of being transformed. Cognitive agents can only act on the known actual variety $a \in \mathbb{D}_A^K \subset A$ for destruction and potentialization. Every decision-choice activity is an action against the existing variety and in favor of a potential variety, where the actual is transformed to a potential and a potential is transformed as a substitute for the actual on the basis of a real cost-benefit configuration. This is the transformation-substitution principle under cost-benefit rationality.

The transformation-substitution process taking place in the actual-potential polarities becomes increasingly complex for command-control decisions of cognitive agents who work under the prescriptive rules of change over the epistemological space. This complexity is the result of the noise in the source-destination transmission mechanism that creates different forms of fuzzy-stochastic uncertainties that constrain the exactness of the management of the command-control decision-choice processes. The destruction of the actual variety creates a vacuum that must be filled. This vacuum may be filled by many potential varieties that cognitive agents have constrained ability to control the actualization of the desired variety in substitutions. This cognitive deficiency due to a system of noise constraints in the transmission mechanism may lead to the actualization of an undesirable potential variety which has a name called an *unintended consequence*. One thing that must be clear in the transformation-substitution process is that the potential variety among all possible varieties that may be actualized is substantially shaped by the method used in destroying the actual variety since cognitive agents have very limited control over the actual-potential system of substitution and transformation of varieties. Here, the meaning of unintended consequence is simply an undesirable or desirable potential variety that has been actualized outside the preferred set. The undesirable preference varieties come from the un-preferred attainable set while the desired preference variety comes from the unattainable preference set. The unintended consequences in the decision-choice space relate to prior ignorance as well as belong to the surprise space. Their computational processes connect to the relational structure of the possibility and probability spaces in terms of prior and posterior information.

As it has been explained in the previous section, data, knowledge, fact and evidence in support of the command-control system over the epistemological space constitute sequential derivatives from the signal dispositions regarding an actual variety and prescriptive actions on a set of potential varieties where each of the varieties is a candidate as an actualized variety. It is within the dynamics of the actual-potential polarity under the substitution-transformation principle that the subject matters of informatics, data analytics, complexity theory, optimal control theory, game theory, engineering theories, and in fact all organic areas of prescriptive science acquire meaning and content in the cognitive struggle for freedom defined under the ability to create sufficient conditions for transformation-substitution actions on actual-potential varieties. In this process, there is an *inter-categorial prescription* with rules and actions for an emergence of a qualitative variety from one category to another. There is also an intra-categorial prescription with a set of rules and actions for the emergence of a quantitative variety within the same qualitative category.

Chapter 6
The Concepts of Data, Fact and Evidence as a Chain of Conceptual Derivatives from Information

Now let us turn our attention to some important concepts that relate to information and info-dynamics and their respective roles in the system of decision-choice actions for the management of command-control systems within the dynamics of the system of social actual-potential polarities. These concepts are data, fact and evidence which are in most cases used interchangeably in ordinary and scientific discourses and also confused with the information without paying explicative attention to them. In this chapter, the differences and similarities will be established and it will be shown as to how they may be viewed within the general theory of information, info-statics, info-dynamics and communication theory. These concepts are essential in the epistemological space in the process of communication and the understanding of both epistemological and ontological events. Importantly, the concept of data is most of the times confused with the grand concept of information just as information is many a time taken to be knowledge and confused with wisdom. These concepts acquire their general and specific meanings over the epistemological space over which decision-choice agents operate with cognitive limitations in language and processing actions. They are irrelevant over the onto-logical space in which nature operates since information is automatically knowl-edge. The decision-choice actions take place in all endeavors over the cognitive space without exception. Let us keep in mind the stated organic relational roles of matter, energy and information in the universal stem in the creation and destruction of varieties and categories, where the role of information is a direct input into knowledge production and the results of knowledge production are direct inputs into decision-choice activities. Information is, thus an indirect input into decision-choice systems. It is also the case that there is a relational structure of continuum and unity among information, data, fact, evidence and wisdom. It is the misunderstanding of this relational structure of continuum and unity that imposes the epistemic and linguistic confusion on the cognitive agents in both knowledge and decision-choice systems. In this Chapter, a search is being undertaken to

© Springer International Publishing AG 2018
K.K. Dompere, *The Theory of Info-Statics: Conceptual Foundations of Information and Knowledge*, Studies in Systems, Decision and Control 112, DOI 10.1007/978-3-319-61639-1_6

separate these concepts in a manner that preserves their identity, continuity and unity. In other words the concepts are to be explicated for understanding and use through their linguistic varieties

6.1   The Concept of Data and Knowledge as Derivatives of Information

A definition of knowledge with its explication has been given in this monograph and also elsewhere in terms of epistemic continuity and unity [R4.7, R4.10]. The concept of information is viewed as the foundation of language and everything knowable in the dynamics of socio-natural actual-potential polarities in which resides a system of dualities and forces of internal changes. In terms of languages, broadly defined, knowing, teaching and learning, information constitutes the primary category from which all other elements over the epistemological space are simple derivatives. To show how these other concepts are derived from information, it is useful to keep in mind the general information definition (GID) that has been provided in this monograph as *characteristic-signal disposition*. The characteristic-signal disposition presents an important point of entry in finding a general definition of data. The general definition of data must meet certain requirements. These requirements are simply the conditions of epistemic continuum and unity with the general information definition, and yet data retains its concept and identity without losing itself in the general concept of information. The conditions for defining the *concept of data* may be achieved by combining the general information definition (GID) with acquaintance which includes observation, experimentation and others at all levels of informing, knowing, and learning, deciding and choosing. On the basis of the general information definition with its analytical structure and acquaintance, the definition of the concept of data may be offered to answer the question of what is data.

Definition 6.1.1: Data
Data is either a discretization or a finite intervalization of the signal disposition $\mathbb{S}_\ell$ into a recognizable form for recording and construction of databases Δ_{ϕ_ℓ} in relation to a particular phenomenon $\phi_\ell \in \Phi$ with element $\omega_\ell \in \Phi$ for processing and to relate it to the characteristic disposition $\mathbb{X}_\ell$ for the verification of the identity of a variety or categorial variety. If D is a discretization operator then $\mathsf{D}(\mathbb{S}_\ell) = \Delta_{\phi_\ell} = \{\delta_k | k \in \Re_{\phi_\ell}, \phi_\ell \in \Phi \text{ and } \ell \in \mathbb{L}^\infty\}$, δ_k is a discretized data point with $\Re_\phi$ as its index set from the signal disposition and Δ_{ϕ_ℓ} is the collection of all elements that form a discrete data set. On the other hand, if $\mathfrak{L}$ is an intervalization operator then $\mathfrak{L}(\mathbb{S}_\ell) = \Lambda_{\phi_\ell} = \{\mathfrak{A}_k | k \in \mathsf{R}_\phi, \phi_\ell \in \Phi \text{ and } \ell \in \mathbb{L}^\infty\}$, $\mathfrak{A}_k$ is an interval-value data with R_{ϕ_ℓ} as its index set from the signal disposition and Λ_{ϕ_ℓ} is the collection of all elements that form interval-value data set.

Note:

It is possible that $\bigcap_k \delta_k = \emptyset$ under conditions of exactness and $\bigcap_k \delta_k \neq \emptyset$ under the conditions of inexactness. The general defining structure of information is composed of characteristic disposition and signal disposition in the form:

The General Information Definition (GID) $\mathbb{Z}$ is a Cartesian product of the form $\mathbb{X} \otimes \mathbb{S}$ where $\mathbb{X} \neq \emptyset$ is the characteristic disposition (set) and $\mathbb{S} \neq \emptyset$ is the attribute signal disposition (set) given the universal object set $\Omega \neq \emptyset$ and may be written as:

$$\mathbb{Z} = \{(x,s)|x \in \mathbb{X}, s \in \mathbb{S}, \forall \omega \in \Omega, \} = \{\mathbb{X} \otimes \mathbb{S}|\Omega \text{ and } (\mathbb{X} \otimes \mathbb{S}) \neq \emptyset\}$$

where every, $\omega \in \Omega$ is considered as a phenomenon $\phi \in \Phi$ and hence $\Omega = \Phi$. The structure of the general information definition is constrained by *characteristic disposition* of the form $\mathbb{X} = \{(x_1, x_2 \ldots x_j \ldots)| \text{ for any } \omega \in \Omega, \phi \in \Phi \text{ and } j \in \mathbb{J}^\infty\}$ where $\Omega = \Phi$ in the sense that every object, state or process may be considered as a phenomenon. The *total attribute space* is called the *universal characteristic set* $\mathbb{X}$ with x as the generic representation of *attributes* on the basis of which the ontological objects are presented in an infinite set of phenomena. If the elements in the universal object set are naturally formed, defined, identified and separated, then the collection of all x constitutes the characteristic space with infinite elements with an index set of attributes $\mathbb{J}^\infty$. The characteristic set is the characteristic disposition which defines the inner essence of objects in both ontological and epistemological spaces. There is a second constraint which is the *signal disposition* and is written as:

$$\mathbb{S}_\ell = \{(s_{\ell 1}, s_{\ell 2}, \ldots, s_{\ell j}, \ldots)|\omega_\ell \in \Omega, \mathbb{X}_\ell \subset \mathbb{X}, j \in \mathbb{J}_\ell \subset \mathbb{J}^\infty \text{ and } \ell \text{ is fixed in } \mathbb{L}^\infty\},$$

where every object $\omega_\ell \in \Omega$ has a corresponding characteristic disposition $\mathbb{X}_\ell$ and signal disposition $\mathbb{S}_\ell$ in such a way that each $\omega_\ell \in \Omega$ with $\phi_\ell \in \Phi$ is defined by $(\mathbb{X}_\ell \otimes \mathbb{S}_\ell) = \mathbb{Z}_\ell$.

As defined, data is an epistemic construct from the elements of signal disposition through acquaintance and recording. The acquaintance is an epistemic process over the epistemological space. The nature of acquaintance tends to define its validity and degrees of exactness and associated uncertainties and riskiness in all epistemological transformations operating in an actual-potential system of actual-potential polarities. By the definition of data, it is not information but a derivative of information, where the signal disposition constitutes the primary conceptual category and data constitutes the derived conceptual category in the linguistic sequence of informing, knowing, and learning, deciding and choosing. It is through the process of derivatives, that the phenomena of disinformation and misinformation tend to arise.

The accumulated data is simply the accumulation of the elements of signal disposition which may or may not have relational connection to the characteristic disposition of a given phenomenon due to poor acquaintance and interpretations of the encounters. In this respect, it may be useful to see data as represented by vocabularies and information as represented by languages. The concepts of vocabulary and language must be broadly defined to include representations and the corresponding rules of combination of the vocabulary where sound and non-sound, verbal and non-verbal structures are admitted. In this respect and in the process of informing, knowing, and learning, the degree of abstractions in the highest order ranges from knowledge to data and data to information, where information relates to knowledge as its input and information relates to data as an input in the info-dynamics, where signal-disposition acts as the primary conceptual category for data as the derived conceptual category.

These are general definitions which apply to all situations of activities of cognitive agents over the epistemological space with or without intentionality, where decision-choice actions and intentionalities may be related to the dynamics of systems of social actual-potential polarities that generate a system of cost-benefit dualities with dynamics of intentionality. These general definitions also apply to the transforming activities over the ontological space in relation to the natural decision-choice activities of the continual dynamics of systems of actual-potential polarities. It is the relational structure of the characteristic disposition and the signal disposition with the methodology of constructionism-reductionism duality that knowledge is derived from information as input into the general knowledge-production process through the dynamics of acquaintance, epistemics and paradigm of thought, without which nothing is knowable, learnable and teachable. It is useful to keep in mind that under perfect ontological space, information is knowledge and data is indistinguishable from information. This situation of perfect information does not hold over the epistemological space in which cognitive agents operate.

It has been argued that the limitations of cognitive agents and intentionalities of cognitive agents in dealing with the elements of any signal disposition through the mechanism of acquaintance create epistemic uncertainties and *imperfect data structure* composed of defectiveness and deceptiveness over the epistemological space. The epistemic uncertainties and imperfect data structures create important disparities, where the constructs of data structures create knowledge differences between the epistemological and ontological characteristic dispositions. The relational structure of characteristic disposition and signal disposition under the general information definition is applicable in all areas of informing, knowing, learning and decision-choice activities over the epistemological space. In this epistemic framework, *data analysis* is nothing more than the analysis of signal dispositions after discretization and/or intervalization to find either patterns or regularities, and knowledge production is nothing more than the mappings of the results of the data analysis onto the characteristic space to identify either a variety or categorial varieties and their behaviors either in time, over time or both. The size of the data depends on the size of the characteristic disposition and the phenomenon of interest in the form $\omega_\ell \in \Omega$ with $\phi_\ell \in \Phi$, $\forall \ell \in \mathbb{L}^\infty$. Large data-analysis simply mean the analysis of large number of elements in the signal disposition. This is now has a fancy name called big data analytics.

The signal dispositions are surrogate representation of the characteristic dispositions while the characteristic-signal dispositions are the surrogate representations of varieties and categorial varieties which represent the primary objects of knowing by identifying individual identities. It is this representation process that provides an epistemic intentionality for a search of knowledge and allows knowledge to be obtained on specific or general universal elements, states, processes and the nature of socio-natural transformation of micro and macro entities. Nothing is known without the signal disposition, its acquaintance and cognitive capacity of analysis over the epistemological space. The knowledge production, therefore, is a process of identifying a series of surrogates and identifying the connectivity of the series of surrogates to the primary variety or categorial varieties. The identification process is defined in the constructionism-reductionism duality for verification and corroboration between the results of the processed data and the elements of characteristic disposition [R17.15, R17.16]. A definition of knowledge may be offered in terms of the relational structure of signal disposition, characteristic disposition, acquaintance and epistemic processing. A similar definition has been offered in [R4.10].

Definition 6.1.2: Knowledge
Knowledge is the stock of socio-natural primary varieties or categorial varieties in that the concept of data from the signal dispositions have been reasonably identified with the elements of the characteristic dispositions at some justified level of low uncertainty over the epistemological space, where the epistemological characteristic disposition is claimed to be equal to the ontological characteristic disposition under either fuzzy-stochastic, stochastic or fuzzy conditionality in such a way that there is equality between an epistemic variety and an ontological variety of an object, state or process.

This definition inherits the essential properties associate with knowing with the info-dynamics, that is, information dynamics that relate to the disappearance of old varieties and emergence of new varieties. The info-dynamics is always in stock-flow disequilibrium [R17.18]. It always has uncertainty, risk and elements of unsureness as long as the cognitive operations take place over the epistemological space through the time domain. Just as socio-natural transformation is a self-exiting, self-correcting and self-organizing system, the knowledge-production process is also a self-exiting, self-correcting and self-organizing system. Knowledge is derived by processing signal dispositions and relating it to the corresponding elements of characteristic dispositions to reveal the inner essences of varieties in terms of identities, differences and similarities. A similar definition of knowledge is provided in [R4.7, R4.10]. As defined, knowledge is the result of an epistemic process between the signal disposition over the epistemic space and characteristic disposition over the ontological space. In this respect, the establishment of a relation equation for knowledge is at reach. This relational equation must link knowledge to the characteristic disposition of any variety through an epistemic operation on the signal disposition.

Let $\mathbb{K}$ be the knowledge space and E an epistemic operator, then the construct of the knowledge space is an epistemic mapping from the signal dispositions to the characteristic dispositions of varieties and categorial varieties in the form: $\mathbb{K} = \mathsf{E} : \mathbb{S} \to \mathbb{X}$ where $\mathsf{E}(s) = x$ is a complex process that links the elements of signal disposition to the appropriate elements of the characteristic disposition. In other words, the knowledge space may be written as a collection of derived known knowledge items of varieties and categorial varieties that have passed through the possibility and probability spaces with a low level of degree of doubt and a high level of degree of surety. It is useful to keep in mind that the signal disposition and characteristic disposition are set and hence the knowledge space may be symbolically defined and represented as:

$$\mathbb{K} = \{ k_i = (s_i, x_i,) | i \in \mathfrak{I} \subset \mathbb{L}^\infty, \phi \in \Phi, x \in \mathbb{X}, s \in \mathbb{S} \text{ and } \omega \in \Omega \} \qquad (6.1.1)$$

The set of $\mathbb{K}$ is the collection of known varieties $v = (x, \phi, \omega)$ in the knowledge space where $k \in \mathbb{K}$ is a justified item by methodological constructionism-reductionism process under nominalism. It is the stock of social knowledge that has been abstracted from the signal dispositions over the epistemological space with some paradigm of thought. The only thing that is not in here is fuzzy conditionality and stochastic conditionality that will express the distribution of the acceptable degrees of surety. It must be kept in mind that the epistemic process takes place over the epistemological space to create an epistemological variety from the interactions between the space of acquaintance and the space of signal dispositions. The *epistemological variety* is then compared to the *ontological variety* for acceptance of equality with qualifying conditions from the possibility space and probability space. The qualifying conditions enter as conditionality developed from fuzziness and limitationality of the acquaintance and signal dispositions. By analytical comparison of the knowledge definition and data definition, it may be observed that the knowledge is a data-derived construct, where data is signal-derived on the basis of information from the space of varieties.

In the *standard definition of information* (SDI) with conditions of *declarative, objective and semantic* (DOS), the concept of a well-formed and meaningful data is central to the definition in such a way that it is claimed that information cannot be dataless. In this definition of information, what is data and what is the meaning of dataless or absence of data? The meaning of data is absent even though the SDI has data as its central core of meaning. The classification of data into data types, such as primary data, metadata, operational data and derived data, is not helpful in the SDI and semantic theory of information, since the concept of data is not defined within the definitional framework of information. Conditions of data cannot be used to define information. For example, a machine may be derived from steel but a machine cannot be used to define steel. However, by the methodological constructionism-reductionism duality, information about varieties constitutes the primary category of knowing, learning and teaching, while data about varieties constitutes the derived category of knowing, learning and teaching. By the methodological constructionism data is derived from information through the signal disposition and by methodological reductionism data is traced to information through the correspondence with characteristic disposition. How does one relate data to info-dynamics and transformations of categorial varieties? How does the concept of data in SDI help in the understanding of false, true, wrong, right, inexact and inexact information in the decision-choice process in the dynamics of a system of socio-natural polarities? How

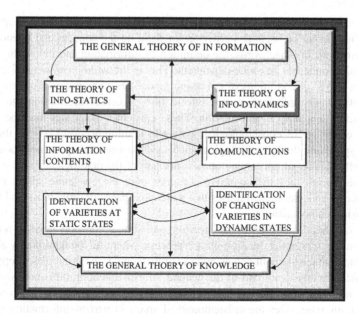

Fig. 6.1 The epistemic geometry of paths of relational structure between the theories of information and knowledge

can one use the conditions of the concept of data to define different logical and mathematical spaces?

The focus of the theory of communication, particularly the mathematical theory of communication has been shifted from its initial intentions to the intentions of the theory of information without the central core of the contents of what is being communicated. The theory of information content that reveals the inner essence of varieties is completely neglected. In this respect, the question of whether information is a property of matter and energy is also put aside. As it has been pointed out, the general theory of information is composed of the theories of info-statics and info-dynamics. Each one of them is made up of a sub-theory of information content and sub-theory of information communication. The concentration on communication is a disservice to the development of the general theory of information on the basis of which knowledge is derived, learning is advanced, and teaching becomes an acceptable social activity for intergenerational transmission of knowledge composed of concepts and ideas (Fig. 6.1).

6.1.1   Epistemic Modalities (Truth, Necessity, Possibility, Surety, Contingency) of Information

When one considers the epistemic conditions in the concepts of falsehood, truth, necessity, possibility and contingency in any information which is defined by the characteristic-signal disposition, the perception of the existence of these epistemic modalities cannot be used as a criterion for defining the concept and phenomenon of information. In other words, the definition of information and the explication of information must not be value-dependent. This is the *value neutrality* of information that allow all conditions of epistemic modalities such as misinformation, disinformation, false information, true information, incorrect information and inexact information to qualify as information. These epistemic modalities must be viewed as part of the set of epistemological noises in communication between the ontological space and the epistemological space and between the epistemological space and itself. This set of epistemic noises is not part of the ontological space that is an epistemic identity for the cognitive activities over the epistemic space. It is the existence of epistemological noises that generates spaces of probability and the risk of decision-choice actions.

The falsehood, truth, necessity, freedom and similar concepts must be related to intentionality within the decision-choice process, where such intentionality also find expressions in cost-benefit games within conflicts under the principles of opposites. This intentionality is not part of the general information definition and must not be part of it. These concepts find their definitions and expressions from information. The intentionality over the epistemological space is part of information whether communicated or not. It must, however, be part of the contents of the information through the signal disposition. The epistemic value nature of information is source-dependent and constitutes part of the system of noises that shapes the information inputs into the knowledge production and its outputs which then become inputs into the decision-choice system to shape the management of the command-control system for the dynamics of the system of actual-potential polarities.

The truth and falsehood of information from the source are nothing more than intentional misrepresentation of some or all the elements of the characteristic disposition through some or all the elements of the signal disposition. In this way, the results of the abstracted data from the signal disposition will create a situation that violates the one-to-one correspondence rule between the characteristics and the signals of the message, as delivered by the contents of the characteristic-signal disposition. The violation of the correspondence rule between the elements of the signal disposition and the elements of characteristic disposition creates an extremely inexact and faulty dataset the processing of which leads to a defective knowledge in the sense that incorrect variety may be decided on as that which corresponds to the signal disposition.

In the whole process of viewing information in relation to matter and energy within the framework of creation and destruction of varieties and categorial varieties, defectiveness, deceptiveness and inexactness cannot be excluded from the definitional structure of information and data that must be seen through the signal disposition. It has been pointed out that these elements are within the set of noises and generate ill-informed conditions. The ill-informed conditions affect the quality of knowing and learning and the decision-choice activities which must be dealt with at the destination over the epistemological space. Here, ill-informed conditions must not be confused with uninformed conditions. The degree of *ill-informedness* will depend on the degree of deception contained in the noise. It is under these ill-informed conditions that ill-informed problem and the construct of corresponding theories arise. The true-false conditions are within the signal disposition in terms of the transmission from the source, and reception, interpretation and recording by the destination. The true-false conditions are applicable to all information transmitted over the epistemological space among cognitive sources and cognitive destinations where the true-false conditions are driven by conditions of intentionalities which are related to cost-benefit configurations in transformation decision-choice actions [R5.13, R5.14]. In other words, the true-false conditions of information are epistemological and not ontological. The degree of deception may also be seen as enhancing vagueness to widen the penumbral region of information-decision interactive processing activities. The process of dealing with the degree of deception gives rise to the need for a non-classical paradigm of thought in information processing to create a crisp knowledge set. The non-classical paradigm is the fuzzy paradigm with fuzzy logic and its mathematics the result of which is always qualified with fuzzy conditionality that helps to capture the subjectivity of interpretative actions [R.4.8, R4.9, R4.30, R4.38].

Alongside of the deceptiveness, there are also conditions of correct and incorrect interpretations of some or all the elements of the signal disposition through the mechanism of acquaintance and in relation to a given object and phenomenon of the form $\omega_\ell \in \Omega$ with $\phi_\ell \in \Phi$ that constitutes a variety. The incorrect interpretations are not due to the presence of deceptiveness, but due to limitations of cognitive agents working in the mechanism to abstract data from the signal disposition, $\mathbb{S}_\ell$ relative to the universal object and phenomenon $\omega_\ell \in \Omega$ with $\phi_\ell \in \Phi$ which specify the variety $v_\ell \in \mathbb{V}$. Deceptive information is a source-dependent. Incorrect interpretation and abstraction from the signal disposition of deceptive information and its use to create faulty data is a destination dependent. The incorrect interpretation and abstraction from the signal disposition may also lead to the development of defectiveness which is composed of limitations on the available space of acquaintance as well as vagueness in the elements in the signal disposition to create faulty and insufficient data. This is a destination-dependent that creates a fuzzy-stochastic dataset, the analysis of which requires the use of the fuzzy paradigm of thought over the fuzzy-stochastic space.

One thing which is clear, is that all these are part of the set of noises which affects the nature of information inputs into the knowledge-production system and hence as derived inputs into the decision-choice process relative to the dynamics of the system of actual-potential polarities with the supporting systems of dualities. The presence of deceptive elements in the source-destination mechanism of transmission of elements of signal disposition presents the epistemological information as existing in true-false duality where every true statement has a false statement support, and every false statement has a true statement support over the channel of communication without which the true and false statements have no epistemic meanings. The acceptance of the truth of any statement is relative to the cognitive computation of the degree of truth constrained by the degree of falsehood within the fuzzy paradigm of thought that helps to resolve the conflict and capture the noise condition in the true-false duality. The signal-disposition of a statement is taken to be true if the fuzzy measure of the degree of truth outweighs the fuzzy measure of the degree of falsehood contained in the same statement with an expressed measure of a *fuzzy conditionality* if the statement is expressed over the fuzzy space or expressed measure of a *fuzzy-stochastic conditionality* if the statement is expressed over the fuzzy-stochastic space [R4.7, R5.34, R5.35, R5.36].

6.1.2   The Concept of Fact as a Derivative from the Concept of Data: The Similarities and Differences in Meanings and Algebraic Structures

The discussions on the concept of data and its analytical structure lead one to consider the epistemic meaning and content of the concept of fact. The epistemic meaning and content of fact will also relate to some important concepts of evidence, empirical information and axiomatic information in the complex system of the process of informing, knowing, learning, deciding and choosing. The definition of data is derived from the signal disposition which constitutes its *primary category* of informing. The definition of the concept of data is derived from the definition of the signal disposition component of information. The definition of the concept of fact must have a primary category of concepts. The candidate to serve as the primary category of the concept of fact is the concept of data in terms of permanency and repetition in the epistemic transformation processes. The definition of the concept of data provides a framework to define the concept of fact. These concepts of data and fact are seen in terms of derivatives of information and hence they must have relational continuum and unity of a conceptual system. These definitions are derived and proceed from the viewpoint of methodological constructionism under the primary category of the concept of information. By the viewpoint of methodological reductionism, these derived concepts must be shown to emerge from the primary in order to maintain the relational continuum with information. The epistemic structure of the methodological constructionism-reductionism duality has been discussed in [R4.10, R4.27]. It is useful to keep in mind that constructionism is methodological process to derive an epistemological characteristic set of a variety, while the reductionism is a methodological process to verify by comparison the equality and difference that reveal a disparity between the epistemological characteristic set and the ontological characteristic set. The disparity between the epistemological characteristic set and the ontological characteristic set is called *epistemic distance*.

Definition 6.1.2.1: Fact
Afact is any permanent and possibly repeated data which is abstracted from the signal disposition that is well-coordinated and confirmed with characteristic disposition of a variety in the past, present or both of a given phenomenon, in the sense

that it belongs to the space of the actual with low degree of doubt which may be produced by quantitative limitation, qualitative limitation or both. Thus if, $D(\mathbb{S}_\ell) = \Delta_{\phi_\ell} = \{\delta_k | k \in \mathfrak{R}_{\phi_\ell}, \phi_\ell \in \Phi$ and $\ell \in \mathbb{L}^\infty\}$ is a discrete dataset and $\mathfrak{F}$ is a fact-conversion operator then $\mathfrak{F}\left(\Delta_{\phi_\ell}\right) = \Theta_{\phi_\ell} = \{f_i = (x_i, \delta_i) | i \in \mathfrak{B}_{\phi_\ell}, \phi_\ell \in \Phi,$ $x \in \mathbb{X}_{\phi_\ell}$ and $\ell \in \mathbb{L}^\infty\}$, where f_i is a fact constructed from the dataset for any given phenomenon $\phi_\ell \in \Phi$ and the corresponding characteristic disposition $\mathbb{X}_{\phi_\ell} \subset \mathbb{X}$ of the universal characteristic set. In the case of continuous data, if $\mathfrak{L}(\mathbb{S}_\ell) = \Lambda_{\phi_\ell} = \{\mathfrak{A}_k | k \in \mathsf{R}_\phi, \phi_\ell \in \Phi$ and $\ell \in \mathbb{L}^\infty\}$ is the continuous data, then the fact-conversion operator works on the continuous data of the form Λ_{ϕ_ℓ}, such that $\mathfrak{F}\left(\Lambda_{\phi_\ell}\right) = \Psi_{\phi_\ell} = \{q_i = (x_i, \mathfrak{A}_i) | i \in \mathsf{R}_{\phi_\ell}, \phi_\ell \in \Phi, x \in \mathbb{X}_{\phi_\ell},$ and $\ell \in \mathbb{L}^\infty\}$, where q_i is a continuous fact that corresponds to the continuous characteristic and data for any given phenomenon $\phi_\ell \in \Phi$ with corresponding object $\omega_\ell \in \Phi$. The term $i \in \mathsf{R}_{\phi_\ell}$ is the index set of facts.

There are some interesting things that must be observed in these definitions. The terms $\mathfrak{F}\left(\Delta_{\phi_\ell}\right)$ and $\mathfrak{F}\left(\Lambda_{\phi_\ell}\right)$ are composite constructs where $\mathfrak{F}\left(\Delta_{\phi_\ell}\right) = \mathfrak{F}(D(\mathbb{S}_\ell))$ and $\mathfrak{F}\left(\Lambda_{\phi_\ell}\right) = \mathfrak{F}(\mathfrak{L}(\mathbb{S}_\ell))$, and by the methodology of reductionism δ_i and $\mathfrak{A}_i$ are the elements of the signal disposition for the phenomenon $\phi_\ell \in \Phi$ and the corresponding object $\omega_\ell \in \Phi$. The abstracted data through the mechanism of acquaintance and its processing into a fact will constitute the cognitive construct of the epistemic-logical signal disposition that must be coordinated with the elements of the characteristic disposition. To the extent to which the abstracted data from the signal disposition marches with the characteristic disposition, a fact is established in relation to a variety $v \in \mathbb{V}$, where $v \in \mathbb{V}\{v\ominus = (\omega, \phi) | \omega \in \Omega, \phi \in \Phi\}$. In other words, the identity of a variety is established by the object $\omega \in \Omega$ and the corresponding phenomenon $\phi \in \Phi$.

Data is either a discretization or a finite intervalization of the signal disposition $\mathbb{S}_\ell$ into recognizable forms for recording and construction of databases Δ_{ϕ_ℓ} in relation to a particular phenomenon with element for processing and to relate it to the characteristic disposition $\mathbb{X}_\ell$ for the verification of the identity of the variety or categorial variety. On the other hand, if $\mathfrak{L}$ is an intervalization operator then $\mathfrak{L}(\mathbb{S}_\ell) = \Lambda_{\phi_\ell} = \{\mathfrak{A}_k | k \in \mathsf{R}_\phi, \phi_\ell \in \Phi$ and $\ell \in \mathbb{L}^\infty\}$, $\mathfrak{A}_k$ is an interval-value data with R_{ϕ_ℓ} as its index set from the signal disposition, and Λ_{ϕ_ℓ} is the collection of all elements that form an interval-value dataset. A fact, therefore, is a relational pair of the form $\left(\mathbb{X}_\ell \otimes \Delta_{\phi_\ell}\right)$ for a discrete dataset and $\left(\mathbb{X}_\ell \otimes \Lambda_{\phi_\ell}\right)$ for a continuous-value dataset. It is important to note that there may be some pairs of the form $(0_i, \delta_i) \in \left(\mathbb{X}_\ell \otimes \Delta_{\phi_\ell}\right)$ for the discrete dataset, and $(0_i, \mathfrak{A}_i) \in \left(\mathbb{X}_\ell \otimes \Lambda_{\phi_\ell}\right)$ for a continuous-value dataset.

A fact about any phenomenon $\phi_\ell \in \Phi$ and any object $\omega_\ell \in \Omega$ for a variety $v \in V$ is data which is abstracted from the signal disposition coordinated with the elements of characteristic disposition in a manner that represents a *low degree of doubt*. The manner of coordination and the low degree of doubt with the elements of characteristic disposition will vary over different phenomenon due to the idea that the true-false elements will vary over different elements of the signal disposition. It is the presence of degrees of doubt associated with any fact as a derivative from data that in the legal cases, circumstantial evidence, witnesses and others that are important elements in the legal decision-choice processes. It is also the same doubt that leads to different explanatory theories about the same phenomenon or prescriptive theories about the same future phenomenon. The same doubt gives the motivation and the logic of repeated experimentation of the process of scientific acquaintance results.

The set of degrees of doubt Q and the set of degrees of surety R may be seen in either non-stochastic-fuzzy space or stochastic fuzzy space. The interactions of the degrees of doubt and surety may be formulated as a fuzzy decision-choice problem for obtaining an optimal degree of doubt with the corresponding degree of surety. The optimal degrees of doubt and surety may then be used to specify the conditions of an acceptance decision. The formulation of the optimal-doubt problem or optimal-surety problem may be done in a way where the fuzzy membership degree of surety is optimized with a constraint of fuzzy membership of degree of doubt. In this respect, one may define $\hat{Q}_\ell = \left\{ \left(\mathbb{X}_\ell \otimes \Delta_{\phi_\ell} \right) | q = (x_i, \delta_i) \right\}$ or $\hat{Q}_\ell = \left\{ \left(\mathbb{X}_\ell \otimes \Lambda_{\phi_\ell} \right) | q = (x_i, \mathfrak{A}_i) \right\}$ with the same index sets as have been specified above.

The membership function of degrees of doubt is specified as $\mu_Q(q) \in (0, 1)$ with $\left(d\mu_Q / dq \right) \leq 0$ which shows a decreasing degree of doubt, and another function of increasing degree of surety $\mu_R(q) \in (0, 1)$ with $\left(d\mu_R / dq \right) \geq 0$ to satisfy the conditions of duality with continuum and unity such that $\mu_R(q) + \mu_Q(q) = 1$ for all q, where Q is the set of degrees of doubt and R is the set of degrees of surety. The variable q may be taken either as a fuzzy non-stochastic (or non-random) variable, or a fuzzy stochastic or fuzzy random variable. This is an optimal-doubt or optimal-surety decision-problem in a fuzzy space the form of which is $D = (Q \cap R)$ where $\left(\mu_R(q) \wedge \mu_Q(q) \right) = \mu_D(q)$. The search for a solution is to optimize $\mu_D(q)$ with respect to q. The solution may be specified as maximize the membership function of the degrees of surety subject to the membership function of the degrees of doubt.

$$\max_q \ \mu_D(q) = \begin{cases} \max_q \mu_R(q) \\ \qquad \text{st.} \ \left[\mu_R(q) - \mu_Q(q) \right] \leq 0 \\ \qquad\qquad \left[\mu_R(q) + \mu_Q(q) \right] = 1 \end{cases} \qquad (6.1.2.1a)$$

$$\max_{q} \; \mu_D(q) = \begin{cases} \min_{q} \mu_Q(q) \\ \quad \text{st.} \; [\mu_Q(q) - \mu_R(q)] \geq 0 \\ \qquad [\mu_R(q) + \mu_Q(q)] = 1 \end{cases} \qquad (6.1.2.1b)$$

The solution to this fuzzy decision problem provides an optimal degree of doubt $q^* = (x_i^*, \delta_i^*)$ or an optimal degree of surety with $q^* = (x_i^*, \mathfrak{A}_i^*)$ (see [R6, R6.3, R6.6, R6.8]. The optimal degree of doubt $\alpha = \mu_Q(q^*)$ with the corresponding optimal degree of surety $\beta = \mu_R(q^*)$ are then used to solve the fact-acceptance decision-choice problem through the use of the method of fuzzy decomposition of the fuzzy sets into rejection and acceptance crisp sets of facts [R4.42, R5.4, R5.13, R5.19, R5.35, R5.37, R5.41]. The acceptable fact is of the form:

$$\left. \begin{array}{l} Q^* = \{q^* = (x^*{}_i, \delta_i^*) | \mu_Q(q^*) \leq \alpha^* = \beta^*\} \text{ less doubt} \\ R^* = \{q^* = (x^*{}_i, \mathfrak{A}_i^*) | \mu_R(q^*) \geq \beta^* = \alpha^*\} \text{ greater surety} \end{array} \right\} \qquad (6.1.2.2)$$

It is useful to always keep in mind the conceptual structure of duality with relational continuum and unity where the optimal degree of surety is equal to the optimal degree of doubt, that is $\alpha^* = \beta^*$ and that $[\mu_Q(q) | q \in Q^*] + [\mu_R(q*) | q \in R^*] = 1$. The optimal degree of doubt or the optimal degree of surety creates fuzzy rationality or fuzzy-stochastic rationality on the basis of which conditions of either non-stochastic fuzzy or fuzzy-stochastic conditionality are developed to qualifying acceptances of solutions to decision-choice problems in fuzzy spaces for what constitutes as fact in support of the identity of a variety.

6.1.3 The Similarities and Differences Between Concepts of Evidence and Facts

Since the concept of fact has been defined and explicated, it is appropriate to turn our attention to examine the relational structure of the concepts of fact and evidence concerning a variety $v \in \mathbb{V}$ in terms of their differences and similarities. The notion of evidence is a disturbing one in general decision-choice process since evidence is a determining factor of the direction of decision-choice action involving true-false, rejection-acceptance claims in all areas of cognitive activities of scientific and non-scientific frames. The concept of evidence also has special and important relation to the cognitive chain of informing, knowing, teaching and learning from the characteristic-signal disposition that forms the primary category in establishing information about ontological and epistemological varieties. It must be stated that the results of observations and experimentations are part of acquaintance and they cannot be taken as facts without processing. They are simply dataset which is discretized or intervalized from the signal decomposition. Evidence has become not only an important but an indispensable part of the decision-choice process as well as established conditions of judgment of the validity of statements, events, assertions and claims of truth, falsehood and knowledge, and counterclaims of reality in theory and practice. The question then is: what is *evidence*? The answer to this question requires a separation between concepts called *conceptual ideas* and things taken to represent concepts called *conceptual things*. In this respect there is the *concept of evidence* and *evidential things*. The definition of the concept of evidence provides the boundaries of things that can satisfy the things which may be used as evidence. In other words, there is an epistemic concept of evidence and pragmatic representation that is induced and justified by the epistemic framework as acceptable in representation.

In this analytical framework, the definition of the concept of evidence indicates the kind of things that will be eligible to meet the defined condition of evidence. The definitional framework deals with the problem of establishing the conditions of eligibility of things evidential for the claims of any given object $\omega_\ell \in \Omega$ with the corresponding phenomenon $\phi_\ell \in \Phi$ establishing a variety $v_\ell \in \mathbb{V}$ whatever the claims may be. The concept of evidence establishes the information and knowledge conditions for justification, corroboration and falsification of a belief in the belief system and a claim in the system of claims. Evidential things are elements presented to meet the evidential condition as established by the definition of the concept of evidence. These evidential things cannot be used to define the concept of evidence. It is the definitional concept of evidence that fixes the conditions of eligibility and acceptability of evidential things. The *evidential things* range over a set of elements and must not be confused with the *concept of evidence*. The concept of evidence, just like the concept of information is general and neither phenomenon-dependent nor object-dependent. Thus there must be a search for the general definition of the concept of evidence.

There is a general concept of evidence derived from the general information definition (GID) of the concept of information. The general definition of the concept of evidence therefore must be information-derived and explicated to serve as defining the boundaries of eligibility and acceptability of different sets of evidential things for different areas of justified and unjustified belief systems. There are no different concepts of evidence to serve as justification of any phenomenon $\phi_\ell \in \Phi$, with the corresponding object $\omega_\ell \in \Omega$ that establishes a variety. There is one and only one concept of evidence. There are, however, different sets of evidential things that can be brought for examination to see whether they meet the eligibility and acceptability conditions for any phenomenon $\phi_\ell \in \Phi$ with a corresponding object $\omega_\ell \in \Omega$. The concept of evidence plays the one of setting the boundaries for judging eligibility and acceptability of the sets of evidential things. These sets of evidential things will vary over different phenomena $\phi_\ell \in \Phi$ with the corresponding object $\omega_\ell \in \Omega$. The variations of the set of evidential-specific things in relation to different phenomenon do not imply that the concept of evidence varies as one examine the elements in the belief system surrounding different phenomena, true-false claims and validity of statements.

When the concept of evidence is distinguished from evidential things, then an epistemic way is open to distinguish different sets of evidential things in terms of degrees of validity and invalidity relative to the same phenomenon $\phi_\ell \in \Phi$ and object $\omega_\ell \in \Omega$. It is the defined concept of evidence that establishes the set of evidential things. Positive and negative sets of evidential things are used to convince others to accept one's belief, claim and assertion, but they are not the defining elements of the concept of evidence. The sets of evidential things are selected to meet the parameters as established by the defining concept of evidence. It is also through the presentation of the set of evidential things, including verbal and non-verbal things, that disinformation and misinformation enter into the source-destination mechanism of messages contained in the signal disposition component of information transmission. The concepts of disinformation and misinformation must be related to the intentionalities of the source (sender), and the acceptability of what is transmitted (signal disposition) must be related to the intentionalities of the destination (the receiver). The formation of intentionalities by both the sender and receiver within the source-destination mechanism will be shaped by awareness-preference conditions and the existing and changing nature of the socio-natural environment. The awareness will depend on the stock-flow conditions of the held knowledge.

There are a number of discussions on evidence and the theory of evidence in claims of knowledge, truth and falsehood of statements, and judgments of validity of current and future events. All these relate to different classifications of evidence such as normative evidence, indicator evidence, confirming evidence in confirmation theory, justifying evidence in justification theory corroborating evidence in corroboration theory in explanatory theories, prescriptive theories and theories of belief systems. Analytically, they all relate to activities of informing, identifying, knowing, deciding and choosing. In other words, evidence' in the final analysis is an information-knowledge phenomenon on the basis of which evidential things are constructed in support of claims and beliefs in phenomenon $\phi_\ell \in \Phi$ with an object $\omega_\ell \in \Omega$. The sets of evidential things are representations that present signal dispositions. On the basis of these, a general definition of the concept of evidence is offered.

Definition 6.1.3.1: Evidence Evidence is an established fact which has been derived from data that has been constructed from the signal disposition of the characteristic-signal disposition for a phenomenon $\phi_\ell \in \Phi$ in relation to an object $\omega_\ell \in \Omega$ and variety $v_\ell \in \mathbb{V}$.

As an establish fact, the concept of evidence meets the defining conditions of data as well as the defining conditions of the signal disposition that contains the message and the characteristic disposition that establishes the conditions of reality. If $\mathbb{E}$ is evidence in support of a phenomenon $\phi_\ell \in \Phi$ and object $\omega_\ell \in \Omega$ then:

$$
\mathbb{E}_{\phi_\ell} = \begin{cases}
\mathfrak{F}\left(\Delta_{\phi_\ell}\right) = \Theta_{\phi_\ell} = \left\{ f_i = (x_i, \delta_i) | i \in \mathfrak{B}_{\phi_\ell}, \phi_\ell \in \Phi, x \in \mathbb{X}_{\phi_\ell}, \ \forall\, \ell \in \mathbb{L}^\infty \right\} \text{discrete evidence} \\
\mathfrak{F}\left(\Lambda_{\phi_\ell}\right) = \Psi_{\phi_\ell} = \left\{ q_i = (x_i, \mathfrak{A}_i) | i \in \mathfrak{R}_{\phi_\ell}, \phi_\ell \in \Phi, x \in \mathbb{X}_{\phi_\ell}, \ \forall\, \ell \in \mathbb{L}^\infty \right\} \text{continuous evidence}
\end{cases}
$$

$$(6.1.3.1)$$

In this system f_i is a discrete variable and q_i is a continuous variable. The definition of the concept of evidence establishes a framework for selecting the sets of *evidential things* $\mathbb{E}_{\phi_\ell}$ in support of claims about $\phi_\ell \in \Phi$ with $\omega_\ell \in \Omega$. While the concept of evidence is neutral to all phenomena, the set of evidential things is phenomenon-specific. It may then be written as:

$$
\mathbb{E}_{\phi_\ell} = \begin{cases}
(e_\lambda, \lambda \in F_{\phi_\ell}) | \mathfrak{F}\left(\Delta_{\phi_\ell}\right) = \Theta_{\phi_\ell} = \left\{ f_i = (x_i, \delta_i) | i \in \mathfrak{B}_{\phi_\ell}, \phi_\ell \in \Phi, x \in \mathbb{X}_{\phi_\ell}, \text{ for } \ell \in \mathbb{L}^\infty \right\} \\
(e_\tau, \tau \in T_{\phi_\ell}) | \mathfrak{F}\left(\Lambda_{\phi_\ell}\right) = \Psi_{\phi_\ell} = \left\{ q_i = (x_i, \mathfrak{A}_i) | i \in \mathfrak{R}_{\phi_\ell}, \phi_\ell \in \Phi, x \in \mathbb{X}_{\phi_\ell}, \text{ for } \ell \in \mathbb{L}^\infty \right\}
\end{cases}
$$

$$(6.1.3.2)$$

The set $\mathsf{E}_{\phi_\ell} = \left(e_\lambda, \ \lambda \in F_{\phi_\ell}\right)$ for the discrete system and the set $\mathsf{E}_{\phi_\ell} = \left(e_\tau, \ \tau \in T_{\phi_\ell}\right)$ for the continuous system are the evidential things that are specific to the phenomenon $\phi_\ell \in \Phi$ and meet the definition of the concept of evidence where F_{ϕ_ℓ} and T_{ϕ_ℓ} are respective index sets. It must be kept in mind that the definition of the *concept of evidence* and the specification of the *evidential things* meet the general conditions of proof in all knowledge systems and all decision-choice environments requiring a show of validity and the demonstration of the nature of variety in the true-false duality.

The sequential structure of definitions and explications offered here is designed to understand the cognitive behavior to the activities over the ontological space relative to the epistemological space in the knowledge-production process in static and dynamic domains. From the ontological information, the epistemological information is constructed from the interactions between the acquaintance and the ontological signal disposition. This is illustrated with epistemic geometry in Fig. 6.2.

The definitional analytics of the sequentially dependent derived concepts of information over the epistemological space is provided in Fig. 6.3. The sequential process is derived from the structure of Fig. 6.2 in terms of observations and abstractions. It is useful to keep in mind that the universe is a family of categories of matter under continual transformation. Matter is a class of elements of sources and destinations for decision-choice actions. Energy is a class of transmitters and receivers between sources and destinations for informing, knowing, learning and teaching. Information is a set of characteristic-signal dispositions that place distinction and similarities on elements to create varieties. This is the knowledge approach to the understanding of information as an input into knowledge production over the epistemological space where the results of knowledge production become inputs into decision-knowledge actions for continual creation of varieties without which changes are unidentifiable and inexplicable. Here, a general

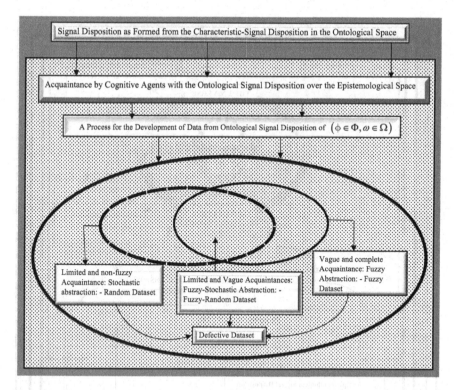

Fig. 6.2 An epistemic geometry of interactions between acquaintance and ontological signal disposition to generate defectiveness in the epistemological information and dataset

framework is provide to the development of informatics and info-dynamics as subject areas of sciences. The subject area of info-dynamics will be dealt with in a separate volume as a follow up of the general theory of information. Every change is a destruction of an existing variety and a creation of a new variety with the loss of the benefit associated with the old variety and the gain of the benefit associated with the new variety. This is the essence of the *asantrofi-anoma principle* of all transformations that has been explained [R17.15, R17.16, R17.18]. The benefit relation of the old-new duality is also the essence of the economic theory of opportunity coast in decision-choice systems and the justification for the study of cost-benefit analysis [R5.13, R5.14]. There is no assessment of the value of the net benefit over the ontological space where the result is continual transformation and production of differentiations while the value of the net benefit over the epistemological space is a subjective assessment by cognitive agents regarding the goodness of the change or the transformation. The general theory of information is not the theory of communication. It is about contents and transmissions of contents among ontological and epistemological objects at any given time point, where the transmission includes the theory of representation such as theories of languages and codes signals as well as the theory of communication over the epistemological space.

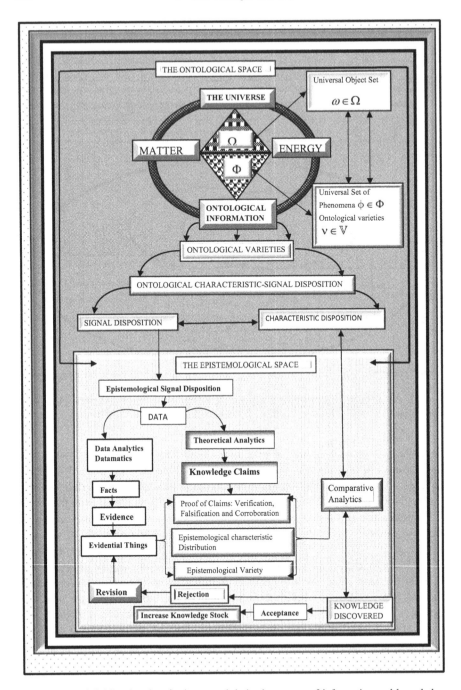

Fig. 6.3 The definitional paths of primary and derived concepts of information and knowledge

6.2 Types of Data, Facts and Evidential Things in Knowing

In this monograph, the general concepts of data, fact and evidence have been defined and explicated from the general information definition of the concept of information. This general information definition is explicated at a general level. It takes on a special set of special characters at different conditions of subject areas at the guidance of the defining and explicating form of information. Different areas of informing, knowing, learning, teaching and deciding will require specific explication that is consistent with the general concepts as have been defined. Different adjectival qualifications of the concept of information will not change the general definitions of information, data, fact and evidence, all of which are derived from the primary category of definition of characteristic-signal information. There are a number of implications that must be noted. Let A be the set of actual and possible areas of knowing with a fuzzy partitioning (overlapping areas) and let $\mathbb{J}$ be its index set of the area of informing, knowing, learning and teaching. Every area of knowing, $\left(A_j,\ j \in \mathbb{J}\right)$ has a set of phenomena $\phi_{\ell j}$ and a set of objects of study $\omega_{\ell j} \in \Omega$ which may be written as:

$$A_j = \left\{ a_{jk} = \left(\phi_{\ell j},\ \omega_{\ell j}\right)_k \middle| \phi_{\ell j} \in \Phi,\ \omega_{\ell j} \in \Omega,\ k \in \mathbb{J}_j\, \forall j \in \mathbb{J}, \forall \ell \in \mathbb{L}^\infty \right\}. \quad (6.2.1)$$

The implication of Eq. (6.2.1) is simply that any information, irrespective of what area of knowing and irrespective of its representation is composed of characteristic disposition and signal disposition to establish varieties where $a_{jk} \in A_j$ represents a variety with $A_j \subset A$ and $A = \bigcup_{j \in \mathbb{J}} A_j$. Furthermore, for each $a_{jk} = \left(\phi_{\ell j},\ \omega_{\ell j}\right)_k$ there is a corresponding characteristic-signal disposition $\mathbb{Z}_{jk}$ that may be written as:

$$\mathbb{Z}_{jk} = \left\{ \left(x_{\ell j},\ s_{\ell j}\right)_k \middle| x_{\ell j} \in \mathbb{X}_k,\ s_{\ell j} \in \mathbb{S}_k, \forall \omega_k \in \Omega, k \in \mathbb{J}_k \right\}$$
$$= \left\{ \left(\mathbb{X}_{\ell j} \otimes \mathbb{S}_{\ell j}\right)_k \middle| \Omega \text{ and } \left(\mathbb{X}_{\ell j} \otimes \mathbb{S}_{\ell j}\right)_k \neq \emptyset \right\} \quad (6.2.2)$$

Any area of knowing deals with a set of actual varieties A_j, a set of pairs of phenomena and objects $\left(\phi_{\ell j} \in \Phi, \omega_{\ell j} \in \Omega\right)$ and a set of characteristic-signal dispositions of the form $\mathbb{Z}_{jk} = \left(x_{\ell j},\ s_{\ell j}\right)_k$ which constitutes the information that provides identities to the varieties. The characteristic-dispositions are the *material essences* of the identities of the varieties, while the signal dispositions are the *informing essence* of the identities of the varieties.

These areas of knowing may be used as adjectival qualification of the type of variety under a search of informing, knowing, learning, teaching, deciding and choosing. These areas allow one to impose a classification of information structures such as *social information, scientific information, engineering information, medical information, legal information* and others. The social information has sub-types of information such as economic information, political information, sociological information, cultural information, religious information and many others. The scientific information has sub-divisions such as chemical information, thermal information, thermodynamic information, astrophysical information, mathematical information, physical information, empirical information, axiomatic information and many others. Engineering information also has sub-types such as electrical-engineering information, chemical-engineering information, biological-engineering information, biomedical-engineering information and many others. Similarly, medical information has sub-types such as surgical information, pharmaceutical information, oncological information, curative information and many others. There are many adjectival qualifications of epistemological information as there are areas of activities of cognitive agents where the sizes of these areas are continually expanding and differentiating in response to the expanding areas of informing and knowing. Keep in mind that the epistemological information has been distinguished from the ontological information and that the distinction finds expressions in the noise space of the source-destination transmission mechanism. Furthermore, the classification of information types and areas of knowledge provides an indisputable justification that information just like matter and energy meets the conditions of variety and identity with defined distribution of characteristic-signal distributions where medical information is distinct from political information and so forth.

Superimposed on all these area-specific types of information is an important conceptual confusion of computer information involving informatics, discrete information, and continuous information, not as operational concepts but as something defining information. Based on this, computational models, computer models, computer abstractions, computational systems and many others have risen, which has given rise to pan-computational systems of reasoning and computer-based *definitionalism* of the phenomenon and content of information. What is missing is defining properties of the concept and meaning of the phenomenon of information. The meaning and the concept of the phenomenon of information is taken as known in all knowledge-search systems. Similarly, the concepts of data, fact, knowledge and evidence are also assumed to be understood without any question to the complete neglect of the computer as an instrument to facilitate the act of processing and transmission over the epistemological space given the paradigm of thought which provides the foundational logic on the basis of which the processing takes place. The paradigms of thought is simply cognitive information processing machines to derive knowledge as inputs into decision-choice systems. At this point, one may be aware that the space of the actual A is always composed of the known actual A^n and the unknown actual A^u such that $A = A^n \cup A^u$ where the expansion of the known actual is a knowledge-process and the expansion of the unknown actual is a variety-transformation process.

6.2.1 The General Information Definition (GID) and the Areas of Knowing as Epistemic Varieties

All of these area-specific types of information cannot be named here, however, a question arises as to what are the differences and similarities. To have some epistemic appreciation and understanding, it must be quickly observed that these information types meet the logical consistency of the general information definition (GID) of characteristic-signal disposition which imposes distinction and identity on varieties. They are all information even though they have adjectival qualifications to create some form of distinction for varieties. All these area-specific types and sub-types of information are varieties and the varieties are under continual transformation that justifies the claim that the dynamics of information is in stock-flow disequilibrium and so also is the dynamics of knowledge production. Every area-type or specific type of information has a corresponding specific type of knowledge. The collection of all these area-specific types of knowledge constitutes the knowledge space. The usefulness of the knowledge space must be related to the decision-choice space in terms of all activities of cognitive agents, where knowledge is transformed into decision and decision is transformed into choice action by the collective or the individual. The transformative power is that the signal disposition becomes an input into the knowledge-production process to result in knowledge. The knowledge becomes an input into the decision process to produce the deliberative result. The deliberative result becomes an input into the choice process with cognitive intentions to produce implementation action which then generates a change in varieties. The structure of the process reveals a continual information production through the matter-energy interplays for continual variety transformation. The corresponding theory in relation to the dynamics of matter-energy-information interplay will be taken up in a separate monograph which will be called the *theory of info-dynamics*.

Epilogue

Reflections on Theories of Information, Knowledge and Decision

The foundational principle behind the analysis and synthesis of the signal disposition is that the creation of data, fact, evidence and acceptance of true-false claims in knowledge production, just like any human activity, is a decision-choice process under the control of cognitive agents. Such a decision-choice process is an input-output process. The inputs are costs and the outputs are benefits, both of which are reversible depending on situations and circumstances in which one finds oneself. The analytical system that has been projected in this monograph is that signal disposition is seen as input and knowledge is seen as output. The movements from information to knowledge are substitution-transformation processes of the signal dispositions that involve the behavior of the dynamics of actual-potential polarity. The basic raw material for the knowledge development in the epistemological space is the characteristic-signal dispositions the cognitive manipulations of which give rise to *fuzziness* and *randomness* in the development and management of human and non-human affairs. The fuzziness expresses itself in the resolution of conflicts in the degrees of accuracy of acquaintance with qualitative disposition while the randomness expresses itself in the resolution of conflicts in the degrees of accuracy of acquaintance with quantitative disposition. The qualitative disposition is seen in vagueness, misinformation, disinformation, approximations and intentionality. The quantitative disposition is seen in terms of volume limitations in acquaintance and approximation. It is the character of disinformation and misinformation in the process of knowing that separates essentially social information and knowledge from scientific information and knowledge through the establishment of the distribution of either the degrees of doubt or surety that may be attached to the results of knowing for acceptance.

Given the organic signal disposition, the problem-solution structures of decision-choice processes depend on the signal disposition and instruments of reasoning which generate an input-output process that defines the success-failure history of the organic knowledge-decision-choice process over the epistemological

© Springer International Publishing AG 2018   157
K.K. Dompere, *The Theory of Info-Statics: Conceptual Foundations
of Information and Knowledge,* Studies in Systems, Decision and Control 112,
DOI 10.1007/978-3-319-61639-1

space composed of the multiplicity of systems dynamics. The individual signal disposition relates to the information-knowledge process of either an individual variety or individual categorial varieties. The nature of the success-failure history is more affected in social information-decision-interactive processes by disinformation and misinformation in the process of knowing than the success-failure history in the natural sciences. This differential effect of disinformation and misinformation is due to social intentionalities in relation to power as expressed in time-dependent preferences, resource distribution and continual creation and transformation of old and new social varieties on the basis of the interactive process of knowledge and decision-choice actions over the epistemological space. Over the ontological space, the knowledge-decision-choice actions that bring about creation, destruction and transformation of natural varieties are under the direct works of nature and out of the control of the existence of cognitive agents, and hence misinformation and disinformation are non-existing in the knowing process. Here, the cognitive agents are themselves the subject of natural transformation without their ability to control the final outcome.

Things are different over the epistemological space, where cognitive agents operate under all kinds of cognitive conditions such as intentionalities, preferences, power acquisitions and many others. Here, the knowledge-decision-choice actions that bring about creation, destruction and transformation of social varieties are under the direct and indirect controls of cognitive agents who must work with the signal dispositions, acquire knowledge, and use it as input for actions to satisfy their intentionalities with individual and collective preferences. The development and uses of misinformation and disinformation to create propaganda are the principal tools of manipulation for advantage in the creation of new social varieties and transformations of old social varieties in the substitution-transformation process of the social actual-potential polarities in the conflict zones of collective preferences. In the substitution-transformation processes, the space of socio-natural knowledge-decision-choice possibilities is just as expansive as the space of the universal transformation possibility of varieties. Every old variety or every new variety is under the information-knowledge principle of stock-flow disequilibrium in destruction-construction processes. Nothing in the universal system can be exempted from this organic process of transformation. A transformation of any element is the destruction an existing variety and the retention of information regarding its previous existence. A transformation of any variety into a new variety is a creation of new information and an increase of the stock of information.

Information-decision-choice-interactive processes through input-output structures of knowledge production and fulfilment of intentionalities are the dominant epistemic culture over the epistemological space. It must be made clear that conceptualization and representation of information are composed of the general definition of the concept and phenomenon which provides a number of advantages in human action. The definition of the concept of information that fixes the boundaries of the phenomenon of information unifies all areas of knowledge under one general logical search process of constructionism-reductionism methodological duality within the framework of theoretical and applied modalities in the social space of actions. Theories and models in social sciences, theories and models in natural and physical sciences, theories and models in biological sciences, theories and models in engineering sciences and many others work with signal dispositions as their inputs of epistemic manipulative processes to derive specific results by the methodological constructionism. Over the epistemological space, these results are the epistemological varieties and categorial varieties whose validities must go through either verification, falsification or corroboration with the elements in the ontological characteristic dispositions by methodological reductionism. It is the organic epistemic process and individual epistemic processes over different areas of cognitive search that provide a constrained picture of the universe at any time, with continual variety transformation dynamics over the infinite horizon of continual information production and storage that form the foundations of knowledge dynamics [R4.13, R8.44, R8.14, R8.16, R8.54]. The explanation of the dynamic picture of information and knowledge may be seen through categorial conversion defining the necessary conditions of transformation, and philosophical Consciencism defining the sufficient conditions of transformation under constructionism-reductionism duality, where every epistemic construction has a corresponding epistemic reduction in relational continuum and unity [R17.15, R17.16].

It is the existence of interactive information-knowledge structures in all areas of informing, knowing, learning and teaching in time and over time that provides us with a framework to divide all theories into two categories of explanatory and prescriptive varieties. The nature of these theories relates to the manner in which the characteristic-signal dispositions are treated under epistemic actions. The focus of the category of explanatory theories is on the destruction of existing varieties and the explanation of the behaviors of existing varieties under their characteristic-signal dispositions. In this respect, any explanatory theory may be devoted to examining inter-categorial behaviors or intra-categorial behaviors or both in all areas of knowing. The focus of the category of prescriptive theories is on the destruction of existing varieties and the creation and transformation of new varieties from the use of characteristic-signal dispositions as inputs into decision-choice systems. The creation and transformation of new varieties may be inter-categorial or intra-categorial or both. The category of prescriptive theories includes theories of engineering and planning and others that act on social actual-potential polarities.

The inter-categorial relations of varieties are qualitative in nature while the intra-categorial relations of varieties are quantitative in nature. For example, a prescriptive theory may be about a transformation of one qualitative variety to another on the basis of their qualitative characteristic-signal dispositions. This is inter-categorial transformation in that there is qualitative movement between two different categorial varieties. A prescriptive theory may be about a transformation of one quantitative variety to another within the same category on the basis of quantitative categorial-signal disposition. This is intra-categorial transformation in that there is quantitative motion between the same varieties in the same category. This explanatory approach defines an integrated epistemic framework of the explanatory process for construction and understanding of the *general theory of unified sciences*, while the prescriptive framework defines an integrated epistemic framework of the prescriptive process for the construction and understanding of the *general theory of unified engineering sciences* in the destruction-construction processes of varieties. It is useful to understand that the general theory of unified sciences with its practices is unbreakably linked to the general theory of engineering sciences and practices over the epistemological space by methodological nominalism in the field of languages that codify and de-codify the signal dispositions from the characteristic dispositions.

The concepts of variety, categorial variety, intra-categorial movement, inter-categorial movement, qualitative characteristic disposition, quantitative characteristic disposition, actual-potential polarity, qualitative disposition, quantitative disposition and others provide foundations for the development of the philosophy and mathematics of information that is not restrictive and closed but open and dynamic, connecting variety-existence to information, to knowledge-development, to decision-choice action, and to creation-destruction action of variety under certain principles of ethics and dynamics of organic and individual quantitative-qualitative dispositions. In this respect, philosophy of information is just as old as philosophy of knowledge whether constructed or not. Every theory of knowledge, irrespective of the area of knowing, has an underlying theory of information or an implicit concept of information on which some epistemic operation is applied to obtain knowledge.

The true-false duality either with excluded middle and disunity or with relational continuum and unity has no existence without information. The philosophy and mathematics of information as is seen in contemporary times must not be restricted to the domain of computational systems and quantitative processing in specific areas such as biomedical information, or informatics and related areas. As has been discussed, data is a derivative from the signal disposition and acquires meaning in the specificity of categorial varieties in the area of investigation as a sub-set of the set of universal varieties. In this respect, the area of informatics is concerned with the science of signal disposition where all types of information systems are sub-derivatives from the signal disposition as the primary element. The differences among the subject areas of knowing reveal themselves as the differences in the defining characteristic dispositions that are subjectively imposed to create varieties of knowing and knowledge areas. In this way different disciplines of research, teaching and knowing are cognitively established for efficiency and specialization.

Critical examination of decision-choice processes, information, knowledge and input-output processes over the epistemological space leads to a number of important questions the answers of which will provide us with the meanings of informing, knowing, learning, teaching, deciding and choosing in the space of human thought and practice under the guidance of a constructed rationality. It has been argued that knowledge is a derivative from information as a primary category of knowing. Informing over the epistemological space is a derivative from the signal disposition. The activities of learning, teaching, deciding and choosing are all at the mercy of the organic and specific information-knowledge processes. Every paradigm of thought is an epistemic information processer under a cognitive action in the search of a variety or categorial varieties to add to the stock of knowledge, and hence every paradigm of thought is useless without information; in fact, without information it is a mirage and vicious in cognition and decision. The development of language is impossible without the existence of ontological varieties the identities of which are defined by their characteristic dispositions and revealed by their signal dispositions. It is the universal existence of ontological varieties that gives meaning to research works in informing, knowing, learning and teaching. This universal existence finds meaning and expression in matter. The changing nature of this universal existence and movement finds expression and meaning in energy as a derivative from matter. It is through the existence of the ontological signal dispositions that the epistemological space is connected to the ontological space by acquaintance.

All definitions in all languages including ordinary and abstract families in theoretical and non-theoretical constructs are about establishing differences, similarities and identities of varieties which is information derived in scientific and non-scientific spaces to give meaning to nominalism in relation to the particular and universal. Vocabulary is about establishing identities and meanings of representation of the phenomena and objects of the form $(\phi \in \Phi, \omega \in \Omega)$ where universality and particularity are in relational continuum and unity of existence to establish varieties. Thus one can distinguish thermodynamics and the corresponding subfields from electrodynamics and corresponding subfields as established by their characteristic-signal dispositions of the nature of their being. How does one know the difference between chemistry and economics, and political science and philosophy? Similarly how does one know that the phenomenon $\phi_1 \in \Phi$ and the corresponding object $\omega_1 \in \Omega$ have been discovered instead of $\phi_2 \in \Phi$ and the corresponding object $\omega_2 \in \Omega$? Their differences and similarities are revealed by the differences of their characteristic-signal dispositions as defining the information structures. It may be kept in mind that by the principle of stock-flow disequilibrium of information, the particular variety finds existence in the universal variety (categorial varieties) and the universal existence of a variety (categorial varieties) finds expression in the particular variety.

Explication in vocabulary is simply to limit the applicable range of meaning as applied to a specific variety and area of thought especially in science and technical areas. Language in all forms is a tool and a vehicle to make the source-destination process of communication of information possible. In the discussions, the conditions on the basis of which information, data, fact, evidence and knowledge are established as well as explained in terms of primary and secondary conceptual derivatives that are intimately connected. It has also been argued that cognitive systems make decisions and choices on the basis of knowledge and not on the basis of information. In this respect, all constructed types of decision-choice rationality are claimed knowledge-supported by defining belief platforms for actions. The transformation decision-choice actions in the ontological space is done by nature itself who uses information because information is knowledge and knowledge is information that requires no processing. The transformation decision-choice actions over the epistemological space is done by cognitive agents with observational limitations and epistemic constraints with which they must process defective information into defective knowledge elements that become inputs into the transformation decision-choice process. It is here that unintended consequences may arise from decision-choice actions. The ontological and epistemological transformation processes have necessary and sufficient conditions. The necessary conditions are established by Categorial Conversion [R17.15] while the sufficient conditions are established by Philosophical Consciencism that provides a guidance to the decision-choice systems with intentionalities [R17.16].

The necessary conditions are external and provide external indication for internal transformation. The sufficient conditions are internal to any variety and indicate the set of internal actions to transform. It is here that a relational structure is established between necessity and freedom, and between freedom and decision-choice actions related to social transformations. In the decision-choice space, necessity defines the cost space and freedom defines the benefit space. At the level of human experience and practice, necessity defines conditions of cost that must be transformed into benefit through freedom of decision-choice actions. Human existence finds meaning in solving the problem of freedom maximization subject to necessity. The actual practice is to maximize benefits subject to costs in variety transformations. The necessity-freedom conditions and the cost-benefit process find expression in the dynamics of the problem-solution process that generate epistemological information in time and over time.

Every decision-choice action is about a resolution of conflicts between the actual and the potential varieties, where the actual variety is defined by the internal real cost-benefit conditions relative to the necessity, and the potential variety is defined by the internal real cost-benefit conditions relative to the internal freedom. The process involves the behavior of information stocks and flows concerning varieties and categorial varieties where a transformation is a transformation of real cost-benefit conditions as revealed by information behavior over time space. The conceptual system for the understanding of this information behavior over time space requires the development of the theory of info-dynamics, the subject matter of which will be dealt with in a follow up monograph. In this possible epistemic development of the theory of info-dynamics, time enters as the fourth dimension in addition to matter, energy and information for the construct of the theory, where the cost-benefit tradeoffs are defined in terms of the opportunity costs of variety transformations [R5.13, R5.14].

The monograph of the theory of info-statics as definitional foundations of information begins with a preface and prologue on the nature of the set of problems of the general theory of information in relation to decision-choice systems. The general theory of information is divided into the theory of info-statics and the theory of info-dynamics. The subject coverage of the theory of info-statics deals with the definition of information to answer the question of what information is and is not for any given time point. The core of this definition is to establish the conditions that allow varieties to be distinguished by difference and varieties to be grouped by similarities and commonness. The difference and similarities are established by partitioned characteristic sets which are essential in decision-choice systems under certainty-uncertainty duality as well as source-destination duality in information transmission and communication. The theory of info-statics ends with an epilogue which is composed of reflections on theories of information, knowledge and decision to provide the required conditions for an entry point into the development of the theory of info-dynamics. The conditions of the theory of info-statics initializes the conditions for the dynamics of the information process. The theory of info-dynamics which is about the production and reproduction of information associated with the transformational dynamics of varieties and categorial varieties will be taken up in a separate monograph.

Multidisciplinary References

R1. Category Theory in Mathematics, Logic and Sciences

[R1.1] Awodey, S., "Structure in Mathematics and Logic: A Categorical Perspective," *Philosophia Mathematica,* Vol. 3. 1996, pp. 209–237.

[R1.2] Bell, J. L., "Category Theory and the Foundations of Mathematics," British *Journal of Science,* Vol. 32, 1981, pp. 349–358.

[R1.3] Bell, J. L., "Categories, Toposes and Sets," *Syntheses,* Vol. 51, 1982, pp. 393–337.

[R1.4] Black, M., *The Nature of Mathematics,* Totowa, N.J., Littlefield, Adams and Co., 1965.

[R1.5] Blass, A., "The Interaction between Category and Set Theory," *Mathematical Applications of Category Theory,* Vol. 30, 1984, pp. 5–29.

[R1.6] Brown, B. and J Woods (eds.), *Logical Consequence; Rival Approaches and New Studies in exact Philosophy: Logic, Mathematics and Science,* Vol. II Oxford, Hermes, 2000.

[R1.7] Domany, J. L., et al., *Models of Neural Networks III,* New York, Springer, 1996.

[R1.8] Feferman, S., "Categorical Foundations and Foundations of Category Theory," in R. Butts (ed.), *Logic, Foundations of Mathematics and Computability,* Boston, Mass., Reidel, 1977, pp. 149–169.

[R1.9] Glimcher, P. W., *Decisions, Uncertainty, and the Brain: The Science of Neoroeconomics,* Cambridge, Mass., MIT Press, 2004.

[R1.10] Gray, J.W. (ed.) *Mathematical Applications of Category Theory* (American Mathematical Society Meeting 89th Denver Colo. 1983)., Providence, R.I., American Mathematical Society, 1984.

[R.1.11] Johansson, Ingvar, *Ontological Investigations: An Inquiry into the Categories of Nature, Man, and Society,* New York, Routledge, 1989.

[R1.12] Kamps, K. H., D. Pumplun, and W. Tholen (eds.) *Category Theory: Proceedings of the International Conference,* Gummersbach, July 6–10, New York, Springer, 1982.

[R1.13] Landry, E., Category Theory: the Language of Mathematics," *Philosophy of Science,* Vol. 66, (Supplement), S14–S27.

[R1.14] Landry E. and J.P Marquis, "Categories in Context: Historical, Foundational and Philosophical," *Philosophia Mathematica,* Vol. 13, 2005, pp. 1–43.

[R1.15] Marquis, J. –P., "Three Kinds of Universals in Mathematics," in B. Brown, and J. Woods (eds.), *Logical Consequence; Rival Approaches and New Studies in exact Philosophy: Logic, Mathematics and Science,* Vol. II Oxford, Hermes, 2000, pp. 191–212.

[R1.16] McLarty, C., "Category Theory in Real Time," *Philosophia Mathematica,* Vol. 2, 1994, pp. 36–44.

[R1.17] McLarty, C., "Learning from Questions on Categorical Foundations," *Philosophia Mathematica,* Vol. 13, 2005, pp. 44–60.

[R1.18] Ross, Don, *Economic theory and Cognitive Science; Microexplanation,* Cambridge, Mass., MIT Press, 2005.

© Springer International Publishing AG 2018

K.K. Dompere, *The Theory of Info-Statics: Conceptual Foundations of Information and Knowledge,* Studies in Systems, Decision and Control 112, DOI 10.1007/978-3-319-61639-1

[R1.19] Rodabaugh, S. et. al., (eds.), *Application of Category Theory to Fuzzy Subsets*, Boston, Mass., Kluwer 1992.

[R1.20] Sieradski, Allan, J., An Introduction to Topology and Homotopy, PWS-KENT Pub. Boston, 1992.

[R1.20] Taylor, J.G. (ed.), *Mathematical Approaches to Neural Networks*, New York North-Holland, 1993.

[R1.21] Van Benthem, J. et al. (eds.), *The Age of Alternative Logics: Assessing Philosophy of Logic and Mathematics Today*, New York, Springer, 2006.

R2. Concepts of Information, Fuzzy Probability, Fuzzy Random Variable and Random Fuzzy Variable

[R2.1] Bandemer, H., "From Fuzzy Data to Functional Relations," *Mathematical Modelling*, Vol. 6, 1987, pp. 419–426.

[R2.2] Bandemer, H. et. al., *Fuzzy Data Analysis*, Boston, Mass, Kluwer, 1992.

[R2.3] Kruse, R. et. al., *Statistics with Vague Data,* Dordrecht, D. Reidel Pub. Co., 1987.

[R2.4] El Rayes, A.B. et. al., "Generalized Possibility Measures," *Information Sciences,* Vol. 79, 1994, pp. 201–222.

[R2.5] Dumitrescu, D., "Entropy of a Fuzzy Process," *Fuzzy Sets and Systems*, Vol. 55, #2, 1993, pp. 169–177.

[R2.6] Delgado, M. et. al., "On the Concept of Possibility-Probability Consistency," *Fuzzy Sets and Systems*, Vol. 21, #3, 1987, pp. 311–318.

[R2.7] Devi, B.B. et. al., "Estimation of Fuzzy Memberships from Histograms," *Information Sciences*, Vol. 35, #1, 1985, pp. 43–59.

[R2.8] Dubois, D. et. al., "Fuzzy Sets, Probability and Measurement," *European Jour. of Operational Research*, Vol. 40, #2, 1989, pp. 135–154.

[R2.9] Fruhwirth-Schnatter, S., "On Statistical Inference for Fuzzy Data with Applications to Descriptive Statistics," *Fuzzy Sets and Systems*, Vol. 50, #2, 1992, pp. 143–165.

[R2.10] Gaines, B.R., "Fuzzy and Probability Uncertainty logics," *Information and Control*, Vol. 38, #2, 1978, pp. 154–169.

[R2.11] Geer, J.F. et. al., "Discord in Possibility Theory," *International Jour. of General Systems*, Vol. 19, 1991, pp. 119–132.

[R2.12] Geer, J.F. et. al., "A Mathematical Analysis of Information-Processing Transformation Between Probabilistic and Possibilistic Formulation of Uncertainty," *International Jour. of General Systems*, Vol. 20, #2, 1992, pp. 14–176.

[R2.13] Goodman, I.R. et. al., *Uncertainty Models for Knowledge Based Systems*, New York, North-Holland, 1985.

[R2.14] Grabish, M. et. al., *Fundamentals of Uncertainty Calculi with Application to Fuzzy Systems*, Boston, Mass., Kluwer, 1994.

[R2.15] Guan, J.W. et. al., *Evidence Theory and Its Applications*, Vol. 1, New York, North-Holland, 1991.

[R2.16] Guan, J.W. et. al., *Evidence Theory and Its Applications*, Vol. 2, New York, North-Holland, 1992.

[R2.17] Hisdal, E., Are Grades of Membership Probabilities?," *Fuzzy Sets and Systems*, Vol. 25, #3, 1988, pp. 349–356.

[R2.18] Höhle Ulrich, "A Mathematical Theory of Uncertainty," in R.R. Yager (ed.) *Fuzzy Set and Possibility Theory: Recent Developments*, New York, Pergamon, 1982, pp. 344–355.

[R2.19] Kacprzyk, Janusz and Mario Fedrizzi (eds.) *Combining Fuzzy Imprecision with Probabilistic Uncertainty in Decision Making*, New York, Plenum Press, 1992.

[R2.20] Kacprzyk, J. et. al., *Combining Fuzzy Imprecision with Probabilistic Uncertainty in Decision Making*, New York, Springer, 1988.

[R2.21] Klir, G.J., "Where Do we Stand on Measures of Uncertainty, Ambignity, Fuzziness and the like?" *Fuzzy Sets and Systems*, Vol. 24, #2, 1987, pp. 141–160.

[R2.22] Klir, G.J. et. al., *Fuzzy Sets, Uncertainty and Information*, Englewood Cliff, Prentice Hll, 1988.

[R2.23] Klir, G.J. et. al., "Probability-Possibility Transformations: A Comparison," *Intern. Jour. of General Systems*, Vol. 21, #3, 1992, pp. 291–310.

[R2.24] Kosko, B., "Fuzziness vs Probability," *Intern. Jour. of General Systems*, Vol. 17, #(1–3) 1990, pp. 211–240.

[R2.26] Manton, K.G. et. al., *Statistical Applications Using Fuzzy Sets*, New York, John Wiley, 1994.

[R2.27] Meier, W., et. al., "Fuzzy Data Analysis: Methods and Industrial Applications," *Fuzzy Sets and Systems*, Vol. 61, #1, 1994, pp. 19–28.

[R2.28] Nakamura, A., et. al., "A logic for Fuzzy Data Analysis," *Fuzzy Sets and Systems*, vol. 39, #2, 1991, pp. 127–132.

[R2.29] Negoita, C.V. et. al., *Simulation, Knowledge-Based Computing and Fuzzy Statistics*, New York, Van Nostrand Reinhold, 1987.

[R2.30] Nguyen, H.T., "Random Sets and Belief Functions," *Jour. of Math. Analysis and Applications*, Vol. 65, #3, 1978, pp. 531–542.

[R2.31] Prade, H. et. al., "Representation and Combination of Uncertainty with belief Functions and Possibility Measures," *Comput. Intell.*, Vol. 4, 1988, pp. 244–264.

[R2.32] Puri, M.L. et. al., "Fuzzy Random Variables," *Jour. of Mathematical Analysis and Applications*, Vol. 114, #2, 1986, pp. 409–422.

[R2.33] Rao, N.B. and A. Rashed, "Some Comments on Fuzzy Random Variables," *Fuzzy Sets and Systems*, Vol. 6, #3, 1981, pp. 285–292.

[R2.34] Sakawa, M. et. al., "Multiobjective Fuzzy linear Regression Analysis for Fuzzy Input-Output Data," *Fuzzy Sets and Systems*, Vol. 47, #2, 1992, pp. 173–182.

[R2.35] Schneider, M. et. al., "Properties of the Fuzzy Expected Values and the Fuzzy Expected Interval," *Fuzzy Sets and Systems*, Vol. 26, #3, 1988, pp. 373–385.

[R2.36] Stein, N.E. and K Talaki, "Convex Fuzzy Random Variables," *Fuzzy Sets and Systems*, Vol. 6, #3, 1981, pp. 271–284.

[R2.37] Sudkamp, T., "On Probability-Possibility Transformations," *Fuzzy Sets and Systems*, Vol. 51, #1, 1992, pp. 73–82.

[R2.38] Walley, P., *Statistical Reasoning with Imprecise Probabilities*, London Chapman and Hall, 1991.

[R2.39] Wang, G.Y. et. al., "The Theory of Fuzzy Stochastic Processes," *Fuzzy Sets and Systems*, Vol. 51, #2 1992, pp. 161–178.

[R2.40] Zadeh, L. A., "Probability Measure of Fuzzy Event," *Jour. of Math Analysis and Applications*, Vol. 23, 1968, pp. 421–427.

R3. Exact Science, Inexact Sciences and Information

[R3.1] Achinstein, P., "The Problem of Theoretical Terms," in Brody, Baruch A. (Ed.) *Reading in the Philosophy of Science*, Englewood Cliffs, NJ., Prentice Hall, 1970.

[R3.2] Amo Afer, A. G., *The Absence of Sensation and the Faculty of Sense in the Human Mind and Their Presence in our Organic and Living Body, Dissertation and Other essays 1727–1749*, Halle Wittenberg, Jena, Martin Luther University Translation, 1968.

[R3.3] Beeson, M. J., *Foundations of Constructive Mathematics*, Berlin/New York, Springer, 1985.

[R3.4] Benacerraf, P., "God, the Devil and Gödel," *Monist*, Vol. 51, 1967, pp. 9–32.

[R3.5] Benecerraf, P and H. Putnam (eds.), *Philosophy of Mathematics: Selected Readings*, Cambridge, Cambridge University Press, 1983.

[R3.6] Black, Max, *The Nature of Mathematics*, Totowa, Littlefield, Adams and Co. 1965.

[R3.7] Blanche, R., *Contemporary Science and Rationalism*, Edinburgh, Oliver and Boyd, 1968.

[R3.8] Blanshard, Brand, *The Nature of Thought*, London Allen and Unwin, 1939.

[R3.9] Blauberg, I. V., V.N. Sadovsky and E.G. Yudin, Systems Theory: Philosophical and Methodological Problems, Moscow, Progress Publishers, 1977.

[R3.10] Braithwaite, R. B., *Scientific Explanation*, Cambridge, Cambridge University Press. 1955.

[R3.11] Brody, Baruch A. (ed.), *Reading in the Philosophy of Science*, Englewood Cliffs, N.J., Prentice Hall, 1970.

[R3.12] Brody, Baruch A., "Confirmation and Explanation," in Brody, Baruch A. (ed.) *Reading in the Philosophy of Science*, Englewood Cliffs, N.J., Prentice Hall, 1970, pp. 410–426.

[R3.13] Brouwer, L.E.J., "Intuitionism and Formalism", *Bull of American Math. Soc.*, Vol. 20, 1913, pp. 81–96.; Also in Benecerraf, P. and H. Putnam (eds.), *Philosophy of Mathematics: Selected Readings*, Cambridge, Cambridge University Press, 1983. pp. 77–89.

[R3.14] Brouwer, L.E.J., "Consciousness, Philosophy, and Mathematics," in Benecerraf, P. and H. Putnam (eds.), *Philosophy of Mathematics: Selected Readings*, Cambridge, Cambridge University Press, 1983, pp. 90–96.

[R3.15] Brouwer, L. E. J., *Collected Works, Vol. 1: Philosophy and Foundations of Mathematics* [A Heyting (ed.)], New York, Elsevier, 1975.

[R3.16] Campbell, Norman R., *What is Science?*, New York, Dover, 1952.

[R3.17] Carnap, R., "Foundations of Logic and Mathematics," in *International Encyclopedia of Unified Science*, Chicago, Univ. of Chicago, 1939, pp. 143–211.

[R3.18] Carnap, Rudolf, "On Inductive Logic," *Philosophy of Science*, Vol. 12, 1945, pp. 72–97.

[R3.19] Carnap, Rudolf, "The Methodological Character of Theoretical Concepts," in Herbert Feigl and M. Scriven (eds.) *Minnesota Studies in the Philosophy of Science, Vol. I*, 1956, pp. 38–76.

[R3.20] Charles, David and Kathleen Lennon (eds.), *Reduction, Explanation, and Realism*, Oxford, Oxford University Press, 1992.

[R3.21] Cohen, Robert S. and Marx W. Wartofsky (eds.), *Methodological and Historical Essays in the Natural and Social Sciences*, Dordrecht, D. Reidel Publishing Co. 1974.

[R3.22] Dalen van, D. (ed.), *Brouwer's Cambridge Lectures on Intuitionism*, Cambridge, Cambridge University Press, 1981.

[R3.23] Davidson, Donald, *Truth and Meaning: Inquiries into Truth and Interpretation*, Oxford, Oxford University Press, 1984.

[R3.24] Davis, M., *Computability and Unsolvability*, New York, McGraw-Hill, 1958.

[R3.25] Denonn. Lester E. (ed.), *The Wit and Wisdom of Bertrand Russell*, Boston, MA., The Beacon Press, 1951.

[R3.26] Dummett, M., "The Philosophical Basis of Intuitionistic Logic," in Benecerraf, P. and H. Putnam (eds.), *Philosophy of Mathematics: Selected Readings*, Cambridge, Cambridge University Press, 1983. pp. 97–129.

[R3.27] Feigl, Herbert and M. Scriven (eds.), *Minnesota Studies in the Philosophy of Science*, Vol. I, 1956.

[R3.28] Feigl, Herbert and M. Scriven (eds.), *Minnesota Studies in the Philosophy of Science*, Vol. II, 1958.

[R3.29] Garfinkel, Alan, *Forms of Explanation: Structures of Inquiry in Social Science*, New Haven, Conn., Yale University Press, 1981.

[R3.30] George, F. H., *Philosophical Foundations of Cybernetics*, Tunbridge Well, Great Britain, 1979.

[R3.31] Gillam, B., "Geometrical Illusions," *Scientific American*, January, 1980, pp. 102–111.

[R3.32] Gödel, Kurt., "What is Cantor's Continuum Problem?" in Benecerraf, P. and H. Putnam (eds.), *Philosophy of Mathematics: Selected Readings*, Cambridge, Cambridge University Press, 1983. pp. 470–486.

[R3.33] Gorsky, D.R., *Definition,* Moscow, Progress Publishers, 1974.

[R3.34] Gray, William and Nicholas D. Rizzo (eds.), *Unity Through Diversity.* New York, Gordon and Breach, 1973.

[R3.35] Hart, W. D. (ed.), *The Philosophy of Mathematics,* Oxford, Oxford University Press, 1996.

[R3.36] Hartkamper, and H. Schmidt, Structure and Approximation in Physical Theories, New York, Plenum Press, 1981.

[R3.37] Hausman, David, M., *The Exact and Separate Science of Economics,* Cambridge, Cambridge University Press, 1992.

[R3.38] Helmer, Olaf and Nicholar Rescher, *On the Epistemology of the Inexact Sciences,* P-1513, Santa Monica, CA, Rand Corporation, October 13, 1958.

[R3.39] Hempel, C. G., "Studies in the Logic of Confirmation," *Mind,* Vol. 54, Part I, 1945, pp. 1–26.

[R3.34] Hempel, Carl G., "The Theoretician's Dilemma," in Herbert Feigl and M. Scriven (eds.) *Minnesota Studies in the Philosophy of Science,* Vol. II, 1958, pp. 37–98.

[R3.35] Hempel, C. G. and P. Oppenheim, "Studies in the Logic of Explanation," *Philosophy of Science,* Vol. 15, 1948, pp. 135–175. [also in Brody, Baruch A. (ed.) *Reading in the Philosophy of Science,* Englewood Cliffs, NJ., Prentice- Hall, 1970, pp. 8–27.

[R3.36] Heyting, A., *Intuitionism: An Introduction,* Amsterdam: North-Holland, 1971.

[R3.37] Hintikka, Jackko (ed.), *The Philosophy of Mathematics,* London, Oxford University Press, 1969.

[R3.38] Hockney D. et al. (eds.), *Contemporary Research in Philosophical Logic and Linguistic Semantics,* Dordrecht-Holland, Reidel Pub., Co. 1975.

[R3.39] Hoyninggen-Huene, Paul and F. M. Wuketits, (eds.), *Reductionism and Systems Theory in the Life Science: Some Problems and Perspectives,* Dordrecht, Kluwer Academic Pub. 1989.

[R3.40] Ilyenkov, E.V., *Dialectical Logic: Essays on Its History and Theory,* Moscow, Progress Publishers, 1977.

[R3.41] Kedrov, B. M., "Toward the Methodological Analysis of Scientific Discovery," *Soviet Studies in Philosophy,* Vol. 11962, pp. 45–65.

[R3.42] Kemeny, John G, and P Oppenheim, "On Reduction," in Brody, Baruch A. (ed.) *Reading in the Philosophy of Science,* Englewood Cliffs, NJ., Prentice- Hall, 1970, 307–318.

[R3.43] Klappholz, K., "Value Judgments of Economics," *British Jour. of Philosophy,* Vol. 15, 1964, pp. 97–114.

[R3.44] Kleene, S.C., "On the Interpretation of Intuitionistic Number Theory," Journal of Symbolic Logic, Vol 10, 1945, pp. 109–124.

[R3.45] Kmita, Jerzy, "The Methodology of Science as a Theoretical Discipline," *Soviet Studies in Philosophy,* Spring, 1974, pp. 38–49.

[R3.46] Krupp, Sherman R.,(ed.), *The Structure of Economic Science,* Englewood Cliff, N. J., Prentice-Hall, 1966.

[R3.47] Kuhn, T., *The Structure of Scientific Revolution,* Chicago, University of Chicago Press, 1970.

[R3.48] Kuhn, Thomas, "The Function of Dogma in Scientific Research," in Brody, Baruch A. (ed.) *Reading in the Philosophy of Science,* Englewood Cliffs, NJ., Prentice-Hall, 1970, pp. 356–374.

[R3.49] Kuhn, Thomas, *The Essential Tension: Selected Studies in Scientific Tradition and Change,* Chicago, University of Chicago Press, 1979.

[R3.50] Lakatos, I. (ed.), *The Problem of Inductive Logic,* Amsterdam, North Holland, 1968.

[R3.51] Lakatos, I., *Proofs and Refutations: The Logic of Mathematical Discovery,* Cambridge, Cambridge University Press, 1976.

[R3.52] Lakatos, I., *Mathematics, Science and Epistemology: Philosophical Papers,* Vol. 2, edited by J. Worrall and G. Currie, Cambridge, Cambridge Univ. Press, 1978.

[R3.53] Lakatos, I., *The Methodology of Scientific Research Programmes,* Vol 1, New York, Cambridge University Press, 1978.

[R3.54] Lakatos, Imre and A. Musgrave (eds.), *Criticism and the Growth of Knowledge*, New York, Cambridge University Press, 1979. Holland, 1979, pp. 153–164.

[R3.55] Lawson, Tony, *Economics and Reality*, New York, Routledge, 1977.

[R3.56] Lenzen, Victor, "Procedures of Empirical Science," in Neurath, Otto et al. (eds.), *International Encyclopedia of Unified Science, Vol. 1–10*, Chicago, University of Chicago Press, 1955, pp. 280–338.

[R3.57] Levi, Isaac, "Must the Scientist make Value Judgments?," in Brody, Baruch A. (Ed.) *Reading in the Philosophy of Science*, Englewood Cliffs, NJ., Prentice- Hall, 1970, pp. 559–570.

[R3.58a] Tse-tung, Mao, *On Practice and Contradiction*, in *Selected works of Mao Tse-tung*, Piking, 1937. Also, London, Revolutions, 2008.

[R3.58b] Lewis, David, *Convention: A Philosophical Study*, Cambridge, Mass., Harvard University Press, 1969.

[R3.59] Mayer, Thomas, *Truth versus Precision in Economics*, London, Edward Elgar, 1993.

[R3.60] Menger, Carl, *Investigations into the Method of the Social Sciences with Special Reference to Economics*, New York, New York University Press, 1985.

[R3.61] Mirowski, Philip (ed.), *The Reconstruction of Economic Theory*, Boston, Mass. Kluwer Nijhoff, 1986.

[R3.62] Mueller, Ian, *Philosophy of Mathematics and Deductive Structure in Euclid's Elements*, Cambridge, Mass., MIT Press, 1981.

[R3.63] Nagel, Ernest, "Review: Karl Niebyl, Modern Mathematics and Some Problems of Quantity, Quality, and Motion in Economic Analysis," *The Journal of Symbolic Logic*, 1940, p. 74.

[R3.64] Nagel, E. et al. (ed.), *Logic, Methodology, and the Philosophy of Science*, Stanford, Stanford University Press, 1962.

[R3.65] Narens, Louis, "A Theory of Belief for Scientific Refutations," *Synthese,* Vol. 145, 2005, pp. 397–423.

[R3.66] Narskii, I. S., "On the Problem of Contradiction in Dialectical Logic," *Soviet Studies in Philosophy,* Vol. vi, #4, pp. 3–10, 1965.

[R3.67] Neurath, Otto et al. (eds.), *International Encyclopedia of Unified Science, Vol. 1–10,* Chicago, University of Chicago Press, 1955.

[R3.68] Neurath Otto, "Unified Science as Encyclopedic," in Neurath, Otto et al. (eds.), *International Encyclopedia of Unified Science, Vol. 1–10,* Chicago, University of Chicago Press, 1955, pp. 1–27.

[R3.69] Planck, Max, *Scientific Autobiography and Other Papers*, Westport, Conn. Greenwood, 1971.

[R3.70] Planck, Max, "The Meaning and Limits of Exact Science," in Max Planck, *Scientific Autobiography and Other Papers,* Westport, Conn. Greenwood, 1971, pp. 80–120.

[R3.71] Polanyi, Michael, "Genius in Science," in Robert S. Cohen, and Marx W. Wartofsky (eds.), *Methodological and Historical Essays in the Natural and Social Sciences*, Dordrecht, D. Reidel Publishing Co. 1974, pp. 57–71.

[R3.72] Popper, Karl, *The Nature of Scientific Discovery*, New York, Harper and Row, 1968.

[R3.73] Putnam, Hilary., "Models and Reality," in Benecerraf, P. and H. Putnam (eds.), *Philosophy of Mathematics: Selected Readings*, Cambridge, Cambridge University Press, 1983. pp. 421–444.

[R3.74] Reise, S., *The Universe of Meaning*, New York, The Philosophical Library, 1953.

[R3.75] Robinson, R., *Definition*, Oxford, clarendon Press, 1950.

[R3.76] Rudner, Richard, "The Scientist qua Scientist Makes Value Judgments," *Philosophy of Science*, Vol. 20, 1953, pp. 1–6.

[R3.77] Russell, B., *Our Knowledge of the External World*, New York, Norton, 1929.

[R3.78] Russell, B., *Human Knowledge, Its Scope and Limits*, London, Allen and Unwin, 1948.

[R3.79] Russell, B., *Logic and Knowledge: Essays 1901–1950,* New York, Capricorn Books, 1971.

[R3.80] Russell, B., *An Inquiry into Meaning and Truth*, New York, Norton, 1940.

[R3.81] Russell, Bertrand, *Introduction to Mathematical Philosophy*, London, George Allen and Unwin, 1919.

[R3.82] Russell, Bertrand, *The Problems of Philosophy*, Oxford, Oxford University Press, 1978.

[R3.83] Rutkevih, M. N., "Evolution, Progress, and the Law of Dialectics," *Soviet Studies in Philosophy*, Vol. IV, #3, pp. 34–43, 1965.

[R3.84] Ruzavin, G. I., "On the Problem of the Interrelations of Modern Formal Logic and Mathematical Logic," *Soviet Studies in Philosophy*, Vol. 3, #1, 1964, pp. 34–44.

[R3.85] Scriven, Michael, "Explanations, Predictions, and Laws," in Brody, Baruch A. (ed.) *Reading in the Philosophy of Science, Englewood Cliffs*, NJ., Prentice- Hall, 1970, pp. 88–104.

[R3.86] Sellars, Wilfrid, The Language of Theories," in Brody, Baruch A. (ed.) *Reading in the Philosophy of Science*, Englewood Cliffs, NJ., Prentice- Hall, 1970, pp. 343–353.

[R3.89] Sterman, John, "The Growth of Knowledge: Testing a Theory of Scientific Revolutions with a Formal Model," *Technological Forecasting and Social Change*, Vol. 28, 1995, pp. 93–122.

[R3.90] Tsereteli, S. B. "On the Concept of Dialectical Logic," *Soviet Studies in Philosophy*, Vol. V, #2, pp. 15–21, 1966.

[R3.91] Tullock, Gordon, *The Organization of Inquiry,* Indianapolis, Indiana, Liberty Fund Inc. 1966.

[R3.92] Van Fraassen, B., *Introduction to Philosophy of Space and Time*, New York, Random House, 1970.

[R3.93] Veldman, W., "A Survey of Intuitionistic Descriptive Set Theory," in P.P. Petkov (ed.), Mathematical Logic: Proceedings of the Heyting Conference, New York, Plenum Press, 1990, pp. 155–174.

[R3.94] Vetrov, A. A., "Mathematical Logic and Modern Formal Logic," *Soviet Studies in Philosophy*, Vol. 3, #1, 1964, pp. 24–33.

[R3.95] von Mises, Ludwig, *Epistemological Problems in Economics*, New York, New York University Press, 1981.

[R3.96] Wang, Hao, *Reflections on Kurt Gödel*, Cambridge, Mass. MIT Press, 1987.

[R3.97] Watkins, J. W. N., "The Paradoxes of Confirmation," in Brody, Baruch A. (ed.) *Reading in the Philosophy of Science*, Englewood Cliffs, NJ., Prentice- Hall, 1970, pp. 433–438.

[R3.98] Whitehead, Alfred North, *Process and Reality*, New York, The Free Press, 1978.

[R3.99] Wittgenstein, Ludwig, *Ttactatus Logico-philosophicus*, Atlantic Highlands, N.J., The Humanities Press Inc.1974.

[R3.100] Woodger, J. H., *The Axiomatic Method in Biology*, Cambridge, Cambridge University Press, 1937.

[R3.101] Zeman, Jiři, "Information, Knowledge and Time," in Libor Kubát, and J. Zeman (eds.), *Entropy and Information in the Physical Sciences*, Amsterdam, Elsevier, 1975, pp. 245–260.

R4. Fuzzy Logic, Information and Knowledge-Production

[R4.1] Baldwin, J.F., "A New Approach to Approximate Reasoning Using a Fuzzy Logic," *Fuzzy Sets and Systems*, Vol. 2, #4, 1979, pp. 309–325.

[R4.2] Baldwin, J.F., "Fuzzy Logic and Fuzzy Reasoning," *Intern. J. Man-Machine Stud.*, Vol. 11, 1979, pp. 465–480.

[R4.3] Baldwin, J.F., "Fuzzy Logic and Its Application to Fuzzy Reasoning," in. M. M. Gupta et al. (eds.), *Advances in Fuzzy Set Theory and Applications*, New York, North-Holland, 1979, pp. 96–115.

[R4.4] Baldwin, J.F. et al., "Fuzzy Relational Inference Language," *Fuzzy Sets and Systems*, Vol. 14, #2, 1984, pp. 155–174.

[R4.5] Baldsin, J. and B.W. Pilsworth, "Axiomatic Approach to Implication For Approximate Reasoning With Fuzzy Logic," *Fuzzy Sets and Systems*, Vol. 3, #2, 1980, pp. 193–219.

[R4.6] Baldwin, J.F. et. al., "The Resolution of Two Paradoxes by Approximate Reasoning Using A Fuzzy Logic," *Synthese*, Vol. 44, 1980, pp. 397–420.

[R4.7] Dompere, K. K., *Fuzzy Rationality: Methodological Critique and Unity of Classical, Bounded and Other Rationalities*, (Studies in Fuzziness and Soft Computing, vol. 235) New York, Springer, 2009.

[R4.8] Dompere Kofi K., *Epistemic Foundations of Fuzziness*, (Studies in Fuzziness and Soft Computing, vol. 236) New York, Springer, 2009.

[R4.9] Dompere Kofi K., *Fuzziness and Approximate Reasoning: Epistemics on Uncertainty, Expectation and Risk in Rational Behavior*, (Studies in Fuzziness and Soft Computing, vol. 237) New York, Springer, 2009.

[R4.10] Dompere, Kofi K., *The Theory of the Knowledge Square: The Fuzzy Rational Foundations of Knowledge-Production Systems*, New York, Springer, 2013.

[R4.11] Dompere, Kofi K., "Cost-Benefit Analysis, Benefit Accounting and Fuzzy Decisions: Part I, Theory", *Fuzzy Sets and Systems*, Vol. 92, 1997, pp. 275–287.

[R4.12] Dompere, Kofi K., "The Theory of Social Cost and Costing For Cost-Benefit Analysis in a Fuzzy Decision Space", Fuzzy Sets and Systems. Vol. 76, 1995, pp. 1–24.

[R4.13] Dompere, Kofi K., *Fuzzy Rational Foundations of Exact and Inexact Sciences*, New York, Springer, 2013.

[R4.14] Gaines, B.R., "Foundations of Fuzzy Reasoning," *Inter. Jour. of Man-Machine Studies*, Vol. 8, 1976, pp. 623–668.

[R4.15] Gaines, B.R., "Foundations of Fuzzy Reasoning," in Gupta, M.M. et. al. (eds.), *Fuzzy Information and Decision Processes*, New York North-Holland, 1982, pp. 19–75.

[R4.16] Gaines, B.R., "Precise Past, Fuzzy Future," *International Journal. Of Man-Machine Studies.*, Vol. 19, #1, 1983, pp. 117–134.

[R4.17] Giles, R., "Lukasiewics Logic and Fuzzy Set Theory," *Intern. J. Man-Machine Stud.*, Vol. 8, 1976, pp. 313–327.

[R4.18] Giles, R., "Formal System for Fuzzy Reasoning," *Fuzzy Sets and Systems*, Vol. 2, #3, 1979, pp. 233–257.

[R4.19] Ginsberg, M. L. (ed.), *Readings in Non-monotonic Reason*, Los Altos, Ca., Morgan Kaufman, 1987.

[R4.20] Goguen, J.A., "The Logic of Inexact Concepts," *Synthese*, Vol. 19, 1969, pp. 325–373.

[R4.21] Gottinger, H.W., "Towards a Fuzzy Reasoning in the Behavioral Science," *Cybernetica*, Vol. 16, #2, 1973, pp. 113–135.

[R4.23] Gupta, M.M. et. al., (eds.), *Approximate Reasoning In Decision Analysis*, North Holland, New York, 1982.

[R4.24] Höhle Ulrich and E.P. Klement, *Non-Classical Logics and their Applications to Fuzzy Subsets: A Handbook of the Mathematical Foundations of Fuzzy Set Theory*, Boston, Mass. Kluwer, 1995.

[R4.25] Kaipov, V., Kh. et. al., "Classification in Fuzzy Environments," in M. M. Gupta et al. (eds.), *Advances in Fuzzy Set Theory and Applications*, New York, North-Holland, 1979, pp. 119–124.

[R4.26] Kaufman, A., "Progress in Modeling of Human Reasoning of Fuzzy Logic" in M. M. Gupta et al. (eds.), *Fuzzy Information and Decision Process*, New York, North-Holland, 1982, pp. 11–17.

[R4.27] Lakoff, G., "Hedges: A Study in Meaning Criteria and the Logic of Fuzzy Concepts," *Jour. Philos. Logic*, Vol. 2, 1973, pp. 458–508.

[R4.28] Lee, R.C.T., "Fuzzy Logic and the Resolution Principle," *Jour. of Assoc. Comput. Mach.*, Vol. 19, 1972, pp. 109–119.

[R4.29] LeFaivre, R.A., "The Representation of Fuzzy Knowledge", *Jour. of Cybernetics*, Vol. 4, 1974, pp. 57–66.

[R4.30] Negoita, C.V., "Representation Theorems for Fuzzy Concepts," *Kybernetes*, Vol. 4, 1975, pp. 169–174.

[R4.31] Nowakowska, M., "Methodological Problems of Measurements of Fuzzy Concepts in Social Sciences", *Behavioral Sciences*, Vol. 22, #2, 1977, pp. 107–115.

[R4.33] Skala, H.J., "On Many-Valued Logics, Fuzzy Sets, Fuzzy Logics and Their Applications," *Fuzzy Sets and Systems*, Vol. 1, #2, 1978, pp. 129–149.

[R4.35] Van Fraassen, B.C., "Comments: Lakoff's Fuzzy Propositional Logic," in D. Hockney et. al., (Eds.), *Contemporary Research in Philosophical Logic and Linguistic Semantics* Holland, Reild, 1975, pp. 273–277.

[R4.36] Yager, R.R. et. al. (eds.), An Introduction to Fuzzy Logic Applications in *Intelligent Systems*, Boston, Mass., Kluwer, 1992.

[R4.38] Zadeh, L.A., "Quantitative Fuzzy Semantics," *Inform. Science*, Vol. 3, 1971, pp. 159–176.

[R4.39] Zadeh, L.A., "A Fuzzy Set Interpretation of Linguistic Hedges," *Jour. Cybernetics*, Vol. 2, 1972, pp. 4–34.

[R4.41] Zadeh, L.A, "The Concept of a Linguistic Variable and Its Application to Approximate Reasoning," in K.S. Fu et. al. (eds.), *Learning Systems and Intelligent Robots*, Plenum Press, New York, 1974, pp. 1–10.

[R4.42] Zadeh, L.A., et. al., (eds.), *Fuzzy Sets and Their Applications to Cognitive and Decision Processes*, New York, Academic Press, 1974.

[R4.43] Zadeh, L.A., "The Birth and Evolution of Fuzzy Logic," *Intern. Jour. of General Systems*, Vol. 17, #(2–3) 1990, pp. 95–105.

R5. Fuzzy Mathematics and Paradigm of Approximate Reasoning Under Conditions of Inexactness and Vagueness

[R5.1] Bellman, R.E., "Mathematics and Human Sciences," in J. Wilkinson et. al. (eds.), *The Dynamic Programming of Human Systems*, New York, MSS Information Corp., 1973, pp. 11–18.

[R5.2] Bellman, R.E and Glertz, M., "On the Analytic Formalism of the Theory of Fuzzy Sets," Information *Science*, Vol. 5, 1973, pp. 149–156.

[R5.3] Butnariu, D., "Fixed Points for Fuzzy Mapping," *Fuzzy Sets and Systems*, Vol. 7, #2, pp. 191–207, 1982.

[R5.4] Butnariu, D., "Decompositions and Range for Additive Fuzzy Measures", *Fuzzy Sets and Systems*, Vol. 10, #2, pp. 135–155, 1983.

[R5.5] Chang, C.L., "Fuzzy Topological Spaces," *J. Math. Anal. and Applications*, Vol. 24, 1968, pp. 182–190.

[R5.6] Chang, S.S.L., "Fuzzy Mathematics, Man and His Environment", *IEEE Transactions on Systems, Man and Cybernetics*, SMC-2 1972, pp. 92–93.

[R5.7] Chang, S.S., "Fixed Point Theorems for Fuzzy Mappings," Fuzzy *Sets and Systems*, Vol. 17, 1985, pp. 181–187.

[R5.8] Chapin, E.W., "An Axiomatization of the Set Theory of Zadeh," *Notices, American Math. Society*, 687-02-4 754, 1971.

[R5.9] Chaudhury, A. K. and P. Das, "Some Results on Fuzzy Topology on Fuzzy Sets," *Fuzzy Sets and Systems*, Vol. 56, 1993, pp. 331–336.

[R5.10] Chitra, H., and P.V. Subrahmanyam, "Fuzzy Sets and Fixed Points," *Jour. of Mathematical Analysis and Application*, Vol. 124, 1987, pp. 584–590.

[R5.11] Czogala, J. et. al., Fuzzy Relation Equations On a Finite Set," *Fuzzy Sets and Systems*, Vol. 7, #1, 1982. pp. 89–101.

[R5.12] DiNola, A. et. al., (eds.), *The Mathematics of Fuzzy Systems*, Koln, Verlag TUV Rheinland, 1986.

[R5.13] Dompere, Kofi K., *Cost-Benefit Analysis and the Theory of Fuzzy Decisions: Identification and Measurement Theory* (Series: Studies in Fuzziness and Soft Computing, Vol. 158), Berlin, Heidelberg, Springer, 2004.

[R5.14] Dompere, Kofi K., *Cost-Benefit Analysis and the Theory of Fuzzy Decisions: Fuzzy Value Theory* (Series: Studies in Fuzziness and Soft Computing, Vol. 160), Berling, Heidelberg, Springer, 2004.

[R5.16] Dubois, D. and H. Prade, *Fuzzy Sets and Systems*, New York, Academic Press, 1980.

[R5.17] Dubois, "Fuzzy Real Algebra: Some Results," *Fuzzy Sets and Systems*, Vol. 2, #4, pp. 327–348, 1979.

[R5.18] Dubois, D. and H. Prade, "Gradual Inference Rules in Approximate Reasoning," *Information Sciences*, Vol. 61(1–2), 1992, pp. 103–122.

[R5.19] Dubois, D. and H. Prade, "On the Combination of Evidence in various Mathematical Frameworks,." In: Flamm. J. and T. Luisi, (eds.), *Reliability Data Collection and Analysis*. Kluwer, Boston, 1992, pp. 213–241.

[R5.20] Dubois, D. and H. Prade, "Fuzzy Sets and Probability: Misunderstanding, Bridges and Gaps." *Proc. Second IEEE Intern. Conf. on Fuzzy Systems*, San Francisco, 1993, pp. 1059–1068.

[R5.21] Dubois, D. and H. Prade [1994], "A Survey of Belief Revision and Updating Rules in Various Uncertainty Models," *Intern. J. of Intelligent Systems*, Vol. 9, #1, pp. 61–100.

[R5.22] Filev, D.P. et. al., "A Generalized Defuzzification Method via Bag Distributions," *Intern. Jour. of Intelligent Systems*, Vol. 6, #7, 1991, pp. 687–697.

[R5.23] Goetschel, R. Jr., et. al., "Topological Properties of Fuzzy Number," *Fuzzy Sets and Systems*, Vol. 10, #1, pp. 87–99, 1983.

[R5.24] Goodman, I.R., "Fuzzy Sets As Equivalence Classes of Random Sets" inYager, R.R. (ed.), *Fuzzy Set and Possibility Theory: Recent Development*, New York, Pergamon Press, 1992, pp. 327–343.

[R5.25] Gupta, M.M. et. al., (eds), *Fuzzy Antomata and Decision Processes*, New York, North-Holland, 1977.

[R5.26] Gupta, M.M. and E. Sanchez (eds.), *Fuzzy Information and Decision Processes*, New York, North-Holland, 1982.

[R5.27] Higashi, M. and G.J. Klir, "On measure of fuzziness and fuzzy complements," *Intern. J. of General Systems*, Vol. 8 #3, 1982, pp. 169–180.

[R5.28] Higashi, M. and G.J. Klir, "Measures of uncertainty and information based on possibility distributions," *International Journal of General Systems*, Vol. 9 #1, 1983, pp. 43–58.

[R5.29] Higashi, M. and G.J. Klir, "On the notion of distance representing information closeness: Possibility and probability distributions," *Intern. J. of General Systems,* Vol. 9 #2, 1983, pp. 103–115.

[R5.30] Higashi, M. and G.J. Klir, "Resolution of finite fuzzy relation equations," *Fuzzy Sets and Systems*, Vol. 13, #1, 1984, pp. 65–82.

[R5.31] Higashi, M. and G.J. Klir, "Identification of fuzzy relation systems," *IEEE Trans. on Systems, Man, and Cybernetics*, Vol. 14 #2, 1984, pp. 349–355.

[R5.32] Jin-wen, Z., "A Unified Treatment of Fuzzy Set Theory and Boolean Valued Set theory: Fuzzy Set Structures and Normal Fuzzy Set Structures," *Jour. Math. Anal. and Applications*, Vol. 76, #1, 1980, pp. 197–301.

[R5.33] Kandel, A. and W.J. Byatt, "Fuzzy Processes," *Fuzzy Sets and Systems*, Vol. 4, #2, 1980, pp. 117–152.

[R5.34] Kaufmann, A. and M.M. Gupta, *Introduction to fuzzy arithmetic: Theory and applications*, New York Van Nostrand Rheinhold,. 1991.

[R5.35] Kaufmann, A., *Introduction to the Theory of Fuzzy Subsets*, Vol. 1, New York, Academic Press, 1975.

[R5.36] Klement, E.P. and W. Schwyhla, "Correspondence Between Fuzzy Measures and Classical Measures," *Fuzzy Sets and Systems,* Vol. 7, #1, 1982. pp. 57–70.

[R5.37] Klir, George and Bo Yuan, *Fuzzy Sets and Fuzzy Logic*, Upper Saddle River, NJ Prentice Hall, 1995.

[R5.38] Kruse, R. et al., *Foundations of Fuzzy Systems*, New York, John Wiley and Sons, 1994.

[R5.37] Lasker, G.E. (ed.), *Applied Systems and Cybernetics, Vol. VI: Fuzzy Sets and Systems*, Pergamon Press, New York, 1981.

[R5.38] Lientz, B.P., "On Time Dependent Fuzzy Sets", *Inform, Science*, Vol. 4, 1972, pp. 367–376.

[R5.39] Lowen, R., "Fuzzy Uniform Spaces," *Jour. Math. Anal. Appl.*, Vol. 82, #21981, pp. 367–376.

[R5.40] Michalek, J., "Fuzzy Topologies," *Kybernetika*, Vol. 11, 1975, pp. 345–354.

[R5.41] Negoita, C.V. et. al., *Applications of Fuzzy Sets to Systems Analysis*, Wiley and Sons, New York, 1975.

[R5.42] Negoita, C.V., "Representation Theorems for Fuzzy Concepts," *Kybernetes*, Vol. 4, 1975, pp. 169–174.

[R5.43] Negoita, C.V. et. al., "On the State Equation of Fuzzy Systems," *Kybernetes*, Vol. 4, 1975, pp. 231–214.

[R5.44] Netto, A.B., "Fuzzy Classes," *Notices, American Mathematical Society*, Vol. 68T-H28, 1968, p. 945.

[R5.45] Pedrycz, W., "Fuzzy Relational Equations with Generalized Connectives and Their Applications," *Fuzzy Sets and Systems*, Vol. 10, #2, 1983, pp. 185–201.

[R5.46] Raha, S. et. al., "Analogy Between Approximate Reasoning and the Method of Interpolation," *Fuzzy Sets and Systems*, Vol. 51, #3, 1992, pp. 259–266.

[R5.47] Ralescu, D., "Toward a General Theory of Fuzzy Variables," *Jour. of Math. Analysis and Applications*, Vol. 86, #1, 1982, pp. 176–193.

[R5.48] Rodabaugh, S.E., "Fuzzy Arithmetic and Fuzzy Topology," in G.E. Lasker, (ed.), *Applied Systems and Cybernetics, Vol. VI: Fuzzy Sets and Systems*, Pergamon Press, New York, 1981, pp. 2803–2807.

[R5.49] Rosenfeld, A., "Fuzzy Groups," *Jour. Math. Anal. Appln.*, Vol. 35, 1971, pp. 512–517.

[R5.50] Ruspini, E.H., "Recent Developments In Mathematical Classification Using Fuzzy Sets," in G.E. Lasker, (ed.), *Applied Systems and Cybernetics, Vol. VI: Fuzzy Sets and Systems*, Pergamon Press, New York, 1981. pp. 2785–2790.

[R5.51] Santos, E.S., "Fuzzy Algorithms," *Inform. and Control*, Vol. 17, 1970, pp. 326–339.

[R5.52] Stein, N.E. and K Talaki, "Convex Fuzzy Random Variables," *Fuzzy Sets and Systems*, Vol. 6, #3, 1981, pp. 271–284.

[R5.53] Triantaphyllon, E. et. al., "The Problem of Determining Membership Values in Fuzzy Sets in Real World Situations," in D.E. Brown et. al. (eds), *Operations Research and Artificial Intelligence: The Integration of Problem-solving Strategies*, Boston, Mass., Kluwer, 1990, pp. 197–214.

[R5.54] Tsichritzis, D., "Participation Measures," *Jour. Math. Anal. and Appln.*, Vol. 36, 1971, pp. 60–72.

[R5.55] Turksens, I.B., "Four Methods of Approximate Reasoning with Interval-Valued Fuzzy Sets," *Intern. Journ. of Approximate Reasoning*, Vol. 3, #2, 1989, pp. 121–142.

[R5.56] Turksen, I.B., "Measurement of Membership Functions and Their Acquisition," *Fuzzy Sets and Systems*, Vol. 40, #1, 1991, pp. 5–38.

[R5.57] Wang, P.P. (ed.), *Advances in Fuzzy Sets, Possibility Theory, and Applications*, New York, Plenum Press, 1983.

[R5.58] Wang, Zhenyuan, and George Klir, *Fuzzy Measure Theory*, New York, Plenum Press, 1992.

[R5.59] Wang, P.Z. et. al. (eds.), *Between Mind and Computer: Fuzzy Science and Engineering*, Singapore, World Scientific Press, 1993.

[R5.60] Wang, S., "Generating Fuzzy Membership Functions: A Monotonic Neural Network Model," *Fuzzy Sets and Systems*, Vol. 61, #1, 1994, pp. 71–82.

[R5.61] Wong, C.K., "Fuzzy Points and Local Properties of Fuzzy Topology," *Jour. Math. Anal. and Appln.*, Vol. 46, 19874, pp. 316–328.

[R5.62] Wong, C.K., "Categories of Fuzzy Sets and Fuzzy Topological Spaces," *Jour. Math. Anal. and Appln.*, Vol. 53, 1976, pp. 704–714.

[R5.62] Yager, R.R. et. al., (Eds.), *Fuzzy Sets, Neural Networks, and Soft Computing*, New York, Nostrand Reinhold, 1994.

[R5.62] Zadeh, L.A., "A Computational Theory of Decompositions," *Intern. Jour. of Intelligent Systems*, Vol. 2, #1, 1987, pp. 39–63.

[R5.63] Zimmerman, H.J., *Fuzzy Set Theory and Its Applications,* Boston, Mass, Kluwer, 1985.

R6. Fuzzy Optimization, Information and Decision-Choice Sciences

[R6.1] Bose, R.K. and Sahani D, "Fuzzy Mappings and Fixed Point Theorems," *Fuzzy Sets and Systems*, Vol. 21, 1987, pp. 53–58.

[R6.2] Butnariu D. "Fixed Points for Fuzzy Mappings," *Fuzzy Sets and Systems*, Vol. 7, 1982, pp. 191–207.

[R6.3] Dompere, Kofi K., "Fuzziness, Rationality, Optimality and Equilibrium in Decision and Economic Theories" in Weldon A. Lodwick and Janusz Kacprzyk (Eds.), *Fuzzy Optimization: Recent Advances and Applications* (Series: Studies in Fuzziness and Soft Computing, Vol. 254), Berlin, Heidelberg, Springer, 2010.

[R6.4] Eaves, B.C., "Computing Kakutani Fixed Points," *Journal of Applied Mathematics*, Vol. 21, 1971, pp. 236–244.

[R6.5] Heilpern, S. "Fuzzy Mappings and Fixed Point Theorem," *Journal of Mathematical Analysis and Applications*, Vol. 83, 1981, pp. 566–569.

[R6.6] Kacprzyk, J. et. al., (eds.), *Optimization Models Using Fuzzy Sets and Possibility Theory*, Boston, Mass., D. Reidel, 1987.

[R6.7] Kaleva, O. "A Note on Fixed Points for Fuzzy Mappings", *Fuzzy Sets and Systems,* Vol. 15, 1985, pp. 99–100.

[R6.8] Lodwick, Weldon A and Janusz Kacprzyk (eds.), *Fuzzy Optimization: Recent Advances and Applications,* (Studies in Fuzziness and Soft Computing, Vol. 254), Berlin Heidelberg, Springer, 2010.

[R6.9] Negoita, C.V., "The Current Interest in Fuzzy Optimization," *Fuzzy Sets and Systems*, Vol. 6, #3, 1981, pp. 261–270.

[R6.10] Negoita, C.V., et. al., "On Fuzzy Environment in Optimization Problems," in J. Rose et. al., (eds.), *Modern Trends in Cybernetics and Systems,* Springer, Berlin, 1977, pp. 13–24.

[R6.11] Zimmerman, H.-J., "Description and Optimization of Fuzzy Systems," *Intern. Jour. Gen. Syst.* Vol. 2, #4, 1975, pp. 209–215.

R7. Ideology, Disinformation, Misinformation and Propaganda

[R7.1] Abercrombie, Nicholas et al., *The Dominant Ideology Thesis*, London, Allen and Unwin, 1980.

[R7.2] Abercrombie, Nicholas, *Class, Structure, and Knowledge: Problems in the Sociology of Knowledge*, New York, New York University Press, 1980.

[R7.3] Aron, Raymond, *The Opium of the Intellectuals*, Lanham, MD, University Press of America, 1985.

[R7.4] Aronowitz, Stanley, *Science as Power: Discourse and Ideology in Modern Society*, Minneapolis, University of Minnesota Press, 1988.

[R7.5] Barinaga, M. and E. Marshall, *Confusion on the Cutting Edge*, Science, Vol. 257, July 1992, pp. 616–625.

[R7.6] Barnett, Ronald, *Beyond All Reason: Living with Ideology in the University*, Philadelphia, PA., Society for Research into Higher Education and Open University Press, 2003.

[R7.7] Barth, Hans, *Truth and Ideology*, Berkeley, University of California Press, 1976.

[R7.8] Basin, Alberto, and Thierry Verdie, "The Economics of Cultural Transmission and the Dynamics of Preferences," *Journal of Economic Theory*, Vol. 97, 2001, pp. 298–319.

[R7.9] Bikhchandani, Sushil et al., "A Theory of Fads, Fashion, Custom, and Cultural Change," *Journal of political Economy*, Vol. 100 1992, pp. 992–1026.

[R7.10] Boyd Robert and Peter J Richerson, *Culture and Evolutionary Process*, Chicago, University of Chicago Press, 1985.

[R7.11] Buczkowski, Piotr and Andrzej Klawiter, *Theories of Ideology and Ideology of Theories*, Amsterdam, Rodopi, 1986.

[R7.12] Chomsky, Norm, *Manufacturing Consent*, New York, Pantheo Press, 1988.

[R7.13] Chomsky, N., *Problem of Knowledge and Freedom*, Glasgow, Collins, 1972.

[R7.14] Cole, Jonathan, R., "Patterns of Intellectual influence in Scientific Research," *Sociology of Education*, Vol. 43, 1968, pp. 377–403.

[R7.15] Cole Jonathan, R. and Stephen Cole, *Social Stratification in Science*, Chicago, University of Chicago Press, 1973.

[R7.16] Debackere, Koenraad and Michael A. Rappa, "Institutioal Variations in Problem Choice and Persistence among Scientists in an Emerging Fields," *Research Policy*, Vol. 23, 1994, pp. 425–441.

[R7.17] Fraser, Colin and George Gaskell (eds.), *The Social Psychological Study of Widespread Beliefs*, Oxford, Clarendon Press, 1990.

[R7.18] Gieryn, Thomas, F. "Problem Retention and Problem Change in Science," *Sociological Inquiry*, Vol. 48, 1978, pp. 96–115.

[R7.19] Harrington, Joseph E. Jr, "The Rigidity of social Systems," *Journal of Political Economy*, Vol. 107, pp. 40–64.

[R7.20] Hinich, Melvin and Michael Munger, *Ideology and the Theory of Political Choice*, Ann Arbor University of Michigan Press, 1994.

[R7.21] Hull, D. L., *Science as a Process: An Evolutionary Account of the Social and Conceptual Development of Science*, Chicago, University of Chicago Press, 1988.

[R7.22] Marx, Karl and Friedrich Engels, *The German Ideology*, New York, International Pub, 1970.

[R7.23] Mészáros, István, *Philosophy, Ideology and Social Science*: Essay in Negation and Affirmation, Brighton, Sussex, Wheatsheaf, 1986.

[R7.24] Mészáros, István, *The Power of Ideology*, New York, New York University Press, 1989.

[R7.25] Newcomb, Theodore M. et. al., *Persistence and Change*, New York, John Wiley, 1967.

[R7.26] Pickering, Andrew, *Science as Practice and Culture*, Chicago, University of Chicago Press, 1992.

[R7.27] Therborn, Göran, *The Ideology of Power and the Power of Ideology*, London, NLB Publications, 1980.

[R7.28] Thompson, Kenneth, *Beliefs and Ideology*, New York, Tavistock Publication, 1986.

[R7.29] Ziman, John, "The Problem of 'Problem Choice'," *Minerva*, Vol. 25, 1987, pp. 92–105.

[R7.30] Ziman, John, *Public Knowledge: An Essay Concerning the Social Dimension of Science*, Cambridge, Cambridge University Press, 1968.

[R7.31] Zuckerman, Hrriet, "Theory Choice and Problem Choice in Science," *Sociological Inquiry*, Vol. 48, 1978, pp. 65–95.

R8. Information, Thought and Knowledge

[R8.1] Aczel, J. and Z. Daroczy, *On Measures of Information and their Characterizations*, New York, Academic Press, 1975.

[R8.2] Afanasyev, Social Information and Regulation of Social Development, Moscow, Progress, 1878.

[R8.3] Anderson, J. R., *The Architecture of Cognition*, Cambridge, Mass., Harvard University Press, 1983.

[R8.4] Angelov, Stefan and Dimitr Georgiev, "The Problem of Human Being in Contemporary Scientic Knowledge," *Soviet Studies in Philosophy*, Summer, 1974, pp. 49–66.

[R8.5] Ash, Robert, *Information Theory*, New York, John Wiley and Sons, 1965.

[R8.6] Bergin, J., "Common Knowledge with Monotone Statistics," *Econometrica*, Vol. 69, 2001, pp. 1315–1332.

[R8.7] Bestougeff, Hélén and Gerard Ligozat, *Logical Tools for Temporal Knowledge Representation*, New York, Ellis Horwood, 1992.

[R8.8] Brillouin, L., *Science and information Theory*, New York, Academic Press, 1962.

[R8.9] Bruner, J. S., et. al., *A Study of Thinking*, New York, Wiley, 1956.

[R8.10] Brunner, K. and A. H. Meltzer (eds.), *Three Aspects of Policy and Policy Making: Knowledge, Data and Institutions*, Carnegie-Rochester Conference Series, Vol. 10, Amsterdam, North-Holland, 1979.

[R8.11] Burks, A. W., *Chance, Cause, Reason: An Inquiry into the Nature of Scientific Evidence*, Chicago, University of Chicago Press, 1977.

[R8.12] Calvert, Randall, *Models of Imperfect Information in Politics*, New York, Hardwood Academic Publishers, 1986.

[R8.13] Cornforth, Maurice, *The Theory of Knowledge*, New York, International Pub. 1972.

[R8.14] Cornforth, Maurice, *The Open Philosophy and the Open Society*, New York, International Pub. 1970.

[R8.15] Coombs, C. H., *A Theory of Data*, New York, Wiley, 1964.

[R8.16] Dretske, Fred. I., *Knowledge and the Flow of Information*, Cambridge, Mass., MIT Press, 1981.

[R8.17] Dreyfus, Hubert L., "A Framework for Misrepresenting Knowledge," in Martin Ringle (ed.) *Philosophical Perspectives in Artificial Intelligence*, Atlantic Highlands, N.J., Humanities press, 1979.

[R8.18] Fagin R. et al., *Reasoning About Knowledge*, Cambridge, Mass, MIT Press, 1995.

[R8.19] Geanakoplos, J., "Common Knowledge," *Journal of Economic Perspectives*," Vol. 6, 1992, pp. 53–82.

[R8.20] George, F. H., *Models of Thinking*, London, Allen and Unwin, 1970.

[R8.21] George, F. H., "Epistemology and the problem of perception," *Mind*, Vol. 66, 1957, pp. 491–506.

[R8.22] Harwood, E. C., *Reconstruction of Economics*, Great Barrington, Mass, American Institute for Economic Research, 1955.

[R8.23] Hintikka, J., *Knowledge and Belief*, Ithaca, N. Y., Cornell University Press, 1962.

[R8.24] Hirshleifer, Jack., "The Private and Social Value of Information and Reward to inventive activity," *American Economic Review*, Vol. 61, 1971, pp. 561–574.

[R8.25] Kapitsa, P. L., "The Influence of Scientific Ideas on Society," *Soviet Studies in Philosophy*, Fall, 1979, pp. 52–71.

[R8.26] Kedrov, B. M., "The Road to Truth," *Soviet Studies in Philosophy*, Vol. 4, 1965, pp. 3–53.

[R8.27] Klatzky, R. L., *Human Memory: Structure and Processes*, San Francisco, Ca., W. H. Freeman Pub., 1975.

[R8.28] Kreps, David and Robert Wilson, "Reputation and Imperfect Information," *Journal of Economic Theory*, Vol. 27. 1982, pp. 253–279.

[R8.29] Kubát, Libor and J. Zeman (eds.), *Entropy and Information*, Amsterdam, Elsevier, 1975.

[R8.30] Kurcz, G. and W. Shugar et al (eds.), *Knowledge and Language*, Amsterdom, North-Holland, 1986.

[R8.31] Lakemeyer, Gerhard, and Bernhard Nobel (eds.), *Foundations of Knowledge Representation and Reasoning*, Berlin, Springer, 1994.

[R8.32] Lektorskii, V. A., "Principles involved in the Reproduction of Objective in Knowledge,", *Soviet Studies in Philosophy*, Vol. 4, #4, 1967, pp. 11–21.

[R8.33] Levi, I., *The Enterprise of Knowledge*, Cambridge, Mass. MIT Press, 1980.

[R8.34] Levi, Isaac, "Ignorance, Probability and Rational Choice", *Synthese*, Vol. 53, 1982, pp. 387–417.

[R8.35] Levi, Isaac, "Four Types of Ignorance," *Social Science*, Vol. 44, pp. 745–756.

[R8.36] Marschak, Jacob, *Economic Information, Decision and Prediction: Selected Essays*, Vol. II, Part II, Boston, Mass. Dordrecht-Holland, 1974.

[R8.37] Menges, G. (ed.), *Information, Inference and Decision*, D. Reidel Pub., Dordrecht, Holland, 1974.

[R8.38] Michael Masuch and László Pólos (eds.), *Knowledge Representation and Reasoning Under Uncertainty*, New York, Springer, 1994.

[R8.39] Moses, Y. (ed.), *Proceedings of the Fourth Conference of Theoretical Aspects of Reasoning about Knowledge*, San Mateo, Morgan Kaufmann, 1992.

[R8.40] Nielsen, L.T. et al., "Common Knowledge of Aggregation Expectations," *Econometrica*, Vol. 58, 1990, pp. 1235–1239.

[R8.41] Newell, A., *Unified Theories of Cognition*, Cambridge, Mass. Harvard University Press, 1990.

[R8.42] Newell, A., *Human Problem Solving*, Englewood Cliff, N.J. Prentice- Hall, 1972.

[R8.43] Ogden, G. K. and I. A., *The Meaning of Meaning*, New York, Harcourt-Brace Jovanovich, 1923.

[R8.44] Planck, Max, Scientific Autobiography and Other Papers, Westport, Conn., Greenwood, 1968.

[R8.45] Pollock, J., *Knowledge and Justification*, Princeton, Princeton University Press, 1974.

[R8.46] Polanyi, M., *Personal Knowledge*, London, Routledge and Kegan Paul, 1958.

[R8.47] Popper, K. R., *Objective Knowledge*, London, Macmillan, 1949.

[R8.48] Popper, K. R., *Open Society and it Enemies*, Vols. *1 and 2* Princeton, Princeton Univ. Press, 2013.

[R8.49] Popper, K. R., *The Poverty of Historicism* New York, Taylor and Francis, 2002.

[R8.50] Price, H. H., *Thinking and Experience*, London, Hutchinson, 1953.

[R8.51] Putman, H., *Reason, Truth and History*, Cambridge, Cambridge University Press, 1981.

[R8.52] Putman. H., *Realism and Reason*, Cambridge, Cambridge University Press, 1983.

[R8.53] Putman, H., *The Many Faces of Realism*, La Salle, Open Court Publishing Co., 1987.

[R8.54] Russell, B., *Human Knowledge, its Scope and Limits*, London, Allen and Unwin, 1948.

[R8.55] Russell, B., *Our Knowledge of the External World*, New York, Norton, 1929.

[R8.56] Samet, D., "Ignoring Ignorance and Agreeing to Disagree," *Journal of Economic Theory*, Vol. 52, 1990, pp. 190–207.

[R8.57] Schroder, Harold, M. and Peter Suedfeld (eds.), *Personality Theory and Information Processing*, New York, Ronald Pub. 1971.

[R8.58] Searle J., *Minds, Brains and Science*, Cambridge, Mass., Harvard University Press, 1985.

[R8.59] Shin, H., "Logical Structure of Common Knowledge," *Journal of Economic Theory*, Vol. 60, 1993, pp. 1–13.

[R8.60] Simon, H. A., *Models of Thought*, New Haven, Conn., Yale University Press, 1979.

[R8.61] Smithson, M., *Ignorance and Uncertainty, Emerging Paradigms*, New York, Springer, 1989.

[R8.62] Sowa, John F., *Knowledge Representation: Logical, Philosophical, and Computational Foundations*, Pacific Grove, Brooks Pub., 2000.

[R8.63] Stigler, G. J., The Economics of Information," *Journal of Political Economy*, Vol. 69, 1961, pp. 213–225.

[R8.64] Tiukhtin, V. S., "How Reality Can be Reflected in Cognition: Reflection as a Property of All Matter," *Soviet Studies in Philosophy*, Vol. 3 #1, 1964, pp. 3–12.

[R8.65] Tsypkin, Ya Z., *Foundations of the Theory of Learning Systems*, New York, Academic Press, 1973.

[R8.66] Ursul, A. D., "The Problem of the Objectivity of Information," in Libor Kubát, and J. Zeman (eds.), *Entropy and Information*, Amsterdam, Elsevier, 1975. pp. 187–230.

[R8.67] Vardi, M. (ed.), *Proceedings of Second Conference on Theoretical Aspects of Reasoning about Knowledge*, Asiloman, Ca., Los Altos, Ca, Morgan Kaufman, 1988.

[R8.68] Vazquez, Mararita, et al., "Knowledge and Reality: Some Conceptual Issues in System Dynamics Modeling," *Systems Dynamics Review*, Vol. 12, 1996, pp. 21–37.

[R8.69] Zadeh, L. A., "A Theory of Commonsense Knowledge," in Skala, Heinz J. et al., (eds.), *Aspects of Vagueness*, Dordrecht, D. Reidel Co. 1984, pp. 257–295.

[R8.70] Zadeh, L. A., "The Concept of Linguistic Variable and its Application to Approximate reasoning," *Information Science*, Vol. 8, 1975, pp. 199–249 (Also in Vol. 9, pp. 40–80).

R9. Languages and Information

[R9.1] Agha, Agha, *Language and Social Relations,* Cambridge, Cambridge University Press, 2006.

[R9.2] Aitchison, Jean (ed.), *Language Change: Progress or Decay?* Cambridge, New York, Melbourne: Cambridge University Press, 2001.

[R9.3] Anderson, Stephen, *Languages: A Very Short Introduction.* Oxford: Oxford University Press (2012).

[R9.4] Aronoff, Mark and Fudeman, Kirsten, *What is Morphology. New York, John Wiley & Sons,* 2011.

[R9.5] Bauer, Laurie (ed.), *Introducing linguistic morphology* Washington, D.C.: Georgetown University Press, 2003.

[R9.6] Barber Alex and Robert J Stainton *(eds.). Concise Encyclopedia of Philosophy of Language and Linguistics.* New York, Elsevier, 2010.

[R9.7] Brown, Keith; Ogilvie, Sarah, (eds.), *Concise Encyclopedia of Languages of the World,* New York, Elsevier Science, 2008.

[R9.8] Campbell, Lyle (ed.), *Historical Linguistics: an Introduction Cambridge,* MASS, MIT Press, 2004.

[R9.9] Chomsky, Noam, *Syntactic Structures.* The Hague: Mouton, 1957.

[R9.10] Chomsky, Noam, *The Architecture of Language.* Oxford, Oxford University Press, 2000.

[R9.11] Clarke, David S. *Sources of semiotic: readings with commentary from antiquity to the present* Carbondale: Southern Illinois University Press, 1990.

[R9.12] Collinge, N.E.(ed.), *An Encyclopedia of Language.* London: New York: Routledge(1989).

[R9.13] Comrie, Bernard (ed.), *Language universals and linguistic typology: Syntax and morphology,* Oxford, Blackwell, 1989.

[R9.14] Comrie, Bernard (ed.), *The World's Major Languages.* New York: Routledge, 2009.

[R9.15] Coulmas, Florian, *Writing Systems: An Introduction to Their Linguistic Analysis.* Cambridge, Cambridge University Press, 2002.

[R9.16] Croft, William, Cruse, D. Alan, *Cognitive Linguistics.* Cambridge, Cambridge University Press, 2004.

[R9.17] Croft, William, `Typology'. In Mark Aronoff; Janie Rees-Miller(ed.), *The Handbook of Linguistics,* Oxford, Blackwell. pp. 81–105, 2001.

[R9.18] Crystal, David (ed.) *The Cambridge Encyclopedia of Language. Cambridge: Cambridge University Press,* 1997.

[R9.20] Deacon, Terrence, *The Symbolic Species: The Co-evolution of Language and the Brain,* New York: W.W. Norton & Company, 1997.

[R9.21] Devitt, Michael and Sterelny, Kim, *Language and Reality: An Introduction to the Philosophy of Language.* Boston: MIT Press, 1999.

[R9.22] Duranti, Alessandro "Language as Culture in U.S. Anthropology: Three Paradigms" *Current Anthropology* Vol. **44** (3), pp. 323–348, 2003.

[R9.23] Evans, Nicholas and Levinson, Stephen C, "The myth of language universals: Language diversity and its importance for cognitive science," Vol. *32 (5). Behavioral and Brain Sciences.* pp. 429–492, 2009.

[R9.24] Fitch, W. Tecumseh, *The Evolution of Language,* Cambridge, Cambridge University Press, 2010.

[R9.25] Foley, William A., *Anthropological Linguistics: An Introduction,* Oxford, Blackwell, 1997.

[R9.26] Ginsburg, Seymour, *Algebraic and Automata-Theoretic Properties of Formal Languages,* New York, North-Holland, 1973.

[R9.27] Goldsmith, John A, *The Handbook of Phonological Theory: Blackwell Handbooks in Linguistics.* Oxford, Blackwell Publishers, 1995.

[R9.28] Greenberg, Joseph, *Language Universals: With Special Reference to Feature Hierarchies.* The Hague, Mouton & Co., 1966.

[R9.29] Hauser, Marc D.; Chomsky, Noam; Fitch, W. Tecumseh, "The Faculty of Language: What Is It, Who Has It, and How Did It Evolve?" *Science,* 22 **298** (5598), pp. 1569–1579, 2002.

[R9.30] Hörz, Herbert, "Information, Sign, Image," in Libor Kubát and Jiří Zeman (eds.) *Entropy and Information in Science and Philosophy,* New York, Elsevier, 1975.

[R9.31] International Phonetic Association, *Handbook of the International Phonetic Association: A guide to the use of the International Phonetic Alphabet. Cambridge,* Cambridge University Press (1999).

[R9.32] Katzner, Kenneth, *The Languages of the World,* New York: Routledge, 1999.

[R9.33] Labov, William, *Principles of Linguistic Change vol. I Internal Factors,* Oxford, Blackwell, 1994.

[R9.34] Labov, William, *Principles of Linguistic Change vol. II Social Factors,* Oxford, Blackwell, 2001.

[R9.35] Levinson, Stephen C. *Pragmatics.* Cambridge: Cambridge University Press, 1983.

[R9.36] Lewis, M. Paul (ed.), *Ethnologue: Languages of the World,* Dallas, Tex.: SIL International, 2009.

[R9.37] Lyons, John, *Language and Linguistics,* Cambridge, Cambridge University Press, 1981.

[R9.38] MacMahon, April M.S., *Understanding Language Change,* Cambridge, Cambridge University Press, 1994.

[R9.39] Matras, Yaron; Bakker, Peter, (eds). *The Mixed Language Debate: Theoretical and Empirical Advances.* Berlin: Walter de Gruyter, 2003.

[R9.40] Moseley, Christopher (ed), *Atlas of the World's Languages in Danger, Paris: UNESCO Publishing* (2010).

[R9.41] Nerlich, B. "History of pragmatics". In L. Cummings (ed.), *The Pragmatics Encyclopedia.* New York: Routledge. pp. 192–93, 2010.

[R9.42] Newmeyer, Frederick J., *The History of Linguistics.* Linguistic Society of America ,2005.

[R9.43] Newmeyer, Frederick J., *Language Form and Language Function* (PDF). Cambridge, MA: MIT Press, 1998.

[R9.44] Nichols, Johanna, *Linguistic diversity in space and time.* Chicago, University of Chicago Press, 1992.

[R9.45] *Nichols, Johanna. "Functional Theories of Grammar". Annual Review of Anthropology* Vol. **13**: 1984, pp. 7–117.

[R9.46] *Sandler, Wendy; Lillo-Martin, Diane,* "Natural Sign Languages", *In Mark Aronoff; Janie Rees-Miller (eds.). The Handbook of Linguistics.* Oxford, Blackwell. pp. 533–563, 2001.

[R9.47] Swadesh, Morris, *"The phonemic principle", Language,* Vol. **10** (2): (1934), pp. 117–129.

[R9.48] Tomasello, Michael, "The Cultural Roots of Language". In B. Velichkovsky and D. Rumbaugh 9EDS.), *Communicating Meaning: The Evolution and Development of Language.* New York, Psychology Press. pp. 275–308, 1996.

[R9.49] Tomasello, Michael, *Origin of Human Communication.* Cambridge Mass., MIT Press, 2008.

[R9.50] Thomason, Sarah G, *Language Contact – An Introduction,* Edinburgh, Edinburgh University Press, 2001.

[R9.51] Ulbaek, Ib, "The Origin of Language and Cognition", In J. R. Hurford and C. Knight (eds.). *Approaches to the evolution of language.* Cambridge, Cambridge University Press. *1998,* pp. 30–43.

[R9.52] Van Valin, jr, Robert D., "Functional Linguistics," In Mark Aronoff; Janie Rees-Miller (eds.), *The Handbook of Linguistics.* Oxford, Blackwell. (2001). pp. 319–337.

R10. Language, Knowledge-Production Process and Epistemics

[R10.1] Aho, A. V. "Indexed Grammar - An Extension of Context-Free Grammars" *Journal of the Association for Computing Machinery,* Vol. 15, 1968, pp. 647–671.

[R10.2] Black, Max (ed.), *The Importance of Language,* Englewood Cliffs, N.J, Prentice- Hall, 1962.

[R10.3] Carnap, Rudolff, Meaning and Necessity: A Study in Semantics and Modal Logic, Chicago, University of Chicago Press, 1956.

[R10.4] Chomsky, Norm, "Linguistics and Philosophy" in S. Hook (ed.) *Language and Philosophy,* New York, New York University Press, 1968, pp. 51–94.

[R10.5] Chomsky, Norm, *Language and Mind,* New York, Harcourt Brace Jovanovich, 1972.

[R10.6] Cooper, William S., *Foundations of Logico-Linguistics: A Unified Theory of Information, Language and Logic,* Dordrecht, D. Reidel, 1978.

[R10.7] Cresswell, M.J.., *Logics and Languages,* London, Methuen Pub. 1973.

[R10.8] Dilman, Ilham, *Studies in Language and Reason,* Totowa, N.J., Barnes and Nobles, Books, 1981.

[R10.9] Fodor, Jerry A., *The Language and Thought,* New York, Thom as Y. Crowell Co, 1975.

[R10.10] Givon, Talmy, *On Understanding Grammar,* New York, Academic Press, 1979.

[R10.11] Gorsky, D.R., *Definition,* Moscow, Progress Publishers, 1974.

[R10.12] Hintikka, Jaakko, The Game of Language, Dordrecht, D. Reidel Pub. 1983.

[R10.13] Johnson-Lair, Philip N. *Mental Models: Toward Cognitive Science of Language, Inference and Consciousnes*s, Cambridge, Mass, Harvard University Pres, 1983.

[R10.14] Kandel, A., "Codes Over Languages," *IEEE Transactions on Systems Man and Cybernetics,* Vol. 4, 1975, pp. 135–138.

[R10.15] Keenan, Edward L. and Leonard M. Faltz, *Boolean Semantics for Natural Languages,* Dordrecht, D. Reidel Pub., 1985.

[R10.16] Lakoff, G. Linguistics and Natural Logic, *Synthese,* Vol. 22, 1970, pp. 151–271.

[R10.17] Lee, E.T., et. al., "Notes On Fuzzy Languages," *Information Science,* Vol. 1, 1969, pp. 421–434.

[R10.18] Mackey, A. and D. Merrill (eds.) *Issues in the Philosophy of Language,* New Haven, CT. Yale University Press, 1976.

[R10.19] Nagel, T., "Linguistics and Epistemology" in S. Hook(ed.) *Language and Philosophy,* New York, New York University Press, 1969, pp. 180–184.

[R10.20] Pike, Kenneth, *Language in Relation to a Unified Theory of Structure of Human Behavior,* The Hague, Mouton Pub., 1969.

[R10.21] Quine, W.V. O. *Word and object,* Cambridge, Mass, MIT Press, 1960.

[R10.22] Russell, Bernard, *An Inquiry into Meaning and Truth,* Penguin Books, 1970.

[R10.23] Tarski, Alfred, *Logic, Semantics and Matamamethics,* Oxford, Clarendon Press, 1956.

[R10.24] Whorf, B.L. (ed.), *Language, Thought and Reality,* New York, Humanities Press, 1956.

R11. Possible-Actual Worlds and Information Analytics

[R11.1] Adams, Robert M., "Theories of Actuality," *Noûs*, Vol. 8, 1974, pp. 211–231.

[R11.2] Allen, Sture (ed.) *Possible Worlds in Humanities, Arts and Sciences*, Proceedings of Nobel Symposium, Vol. 65, New York, Walter de Gruyter Pub., 1989.

[R11.3] Armstrong, D. M., *A Combinatorial Theory of Possibility*. Cambridge University Press, 1989.

[R11.4] Armstrong, D.M *A World of States of Affairs*, Cambridge, Cambridge University Press, 1997.

[R11.5] Bell, J.S., "Six Possible Worlds of Quantum Mechanics" in Allen, Sture (Ed.) *Possible Worlds in Humanities, Arts and Sciences*, Proceedings of Nobel Symposium, Vol. 65, New York, Walter de Gruyter Pub., 1989. pp. 359–373.

[R11.6] Bigelow, John. "Possible Worlds Foundations for Probability", *Journal of Philosophical Logic*, 5 (1976), pp. 299–320.

[R11.7] Bradley, Reymond and Norman Swartz, *Possible World: An Introduction to Logic and its Philosophy*, Oxford, Bail Blackwell, 1997.

[R11.8] Castañeda, H.-N. "Thinking and the Structure of the World", *Philosophia*, 4 (1974), pp. 3–40.

[R11.9] Chihara, Charles S. *The Worlds of Possibility: Modal Realism and the Semantics of Modal Logic*, Clarendon, 1998.

[R11.10] Chisholm, Roderick. "Identity through Possible Worlds: Some Questions", *Noûs*, 1 (1967), pp. 1–8; reprinted in Loux, *The Possible and the Actual*.

[R11.11] Divers, John, *Possible Worlds*, London: Routledge, 2002.

[R11.12] Forrest, Peter. "Occam's Razor and Possible Worlds", *Monist*, 65 (1982), pp. 456–64.

[R11.13] Forrest, Peter, and Armstrong, D. M. "An Argument Against David Lewis' Theory of Possible Worlds", *Australasian Journal of Philosophy*, 62 (1984), pp. 164–168.

[R11.14] Grim, Patrick, "There is No Set of All Truths", *Analysis*, Vol. 46, 1986, pp. 186–191.

[R11.15] Heller, Mark. "Five Layers of Interpretation for Possible Worlds", *Philosophical Studies*, 90 (1998), pp. 205–214.

[R11.16] Herrick, Paul, *The Many Worlds of Logic*,. Oxford: Oxford University Press, 1999.

[R11.17] Krips, H. "Irreducible Probabilities and Indeterminism", *Journal of Philosophical Logic*, Vol. 18, 1989, pp. 155–172.

[R11.18] Kuhn, Thomas S., "Possible Worlds in History of Science" in Allen, Sture (ed.) *Possible Worlds in Humanities, Arts and Sciences*, Proceedings of Nobel Symposium, Vol. 65, New York, Walter de Gruyter Pub., 1989, pp. 9–41.

[R11.19] Kuratowski, K. and Mostowski, A. *Set Theory: With an Introduction to Descriptive Set Theory*, New York: North-Holland, 1976.

[R11.20] Lewis, David, *On the Plurality of Worlds*, Oxford, Basil Blackwell, 1986.

[R11.21] Loux, Michael J. (ed.) *The Possible and the Actual: Readings in the Metaphysics of Modality*, Ithaca & London: Cornell University Press, 1979.

[R11.22] Parsons, Terence, *Nonexistent Objects*, New Haven, Yale University Press, 1980.

[R11.23] Perry, John, "From Worlds to Situations", *Journal of Philosophical Logic*, Vol. 15, 1986, pp. 83–107.

[R11.24] Rescher, Nicholas and Brandom, Robert. *The Logic of Inconsistency: A Study in Non-Standard Possible-World Semantics And Ontology*, Rowman and Littlefield, 1979.

[R11.25] Skyrms, Brian. "Possible Worlds, Physics and Metaphysics", *Philosophical Studies*, Vol. 30, 1976, pp. 323–32.

[R11.26] Stalmaker, Robert C. "Possible World", *Noûs*, Vol. 10, 1976, pp. 65–75.

[R11.27] Quine, W.V.O. *Word and Object*, M.I.T. Press, 1960.

[R11.28] Quine, W.V.O "Ontological Relativity", *Journal of Philosophy*, 65 (1968), pp. 185–212.

R12. Philosophy of Information and Semantic Information

[R12.1] Aisbett, J., Gibbon, G.: "A practical measure of the information in a logical theory" Journal of Experimental and Theoretical Artificial Intelligence Vol. 11(2), 1999, pp. 201–218.

[R12.2] Badino, M.: "An Application of Information Theory to the Problem of the Scientific Experiment" *Synthese* Vol. 140, 2004, pp. 355–389.

[R12.3] Bar-Hillel, Y. (ed.), *Language and Information: Selected Essays on Their Theory and Application,* Reading, Addison-Wesley, (1964)

[R12.4] Bar-Hillel, Y., Carnap, R. "An Outline of a Theory of Semantic Information," (1953); in Bar-Hillel, Y. (ed.), *Language and Information: Selected Essays on Their Theory and Application,* Reading, Addison-Wesley, (1964), pp. 221–274.

[R12.5] Barwise, J., Seligman, J.: *Information Flow: The Logic of Distributed Systems,* Cambridge, University Press, Cambridge (1997)

[R12.6] Braman, S.: "Defining Information," *Telecommunications Policy,* Vol. 13, pp. 233–242 (1989)

[R12.7] Bremer, M.E.: "Do Logical Truths Carry Information?" *Minds and Machines* Vol. 13(4), 2003, pp. 567–575.

[R12.8] Bremer, M. and Cohnitz, D., *Information and Information Flow: an Introduction,* Ontos Verlag, Frankfurt, Lancaster, 2004.

[R12.9] Chaitin, G.J., *Algorithmic Information Theory,* Cambridge, Cambridge University Press, 1987.

[R12.10] Chalmers, D.J.: *The Conscious Mind: In Search of a Fundamental Theory,* New York, Oxford Univ. Press, (1996).

[R12.11] Cherry, C.: *On Human Communication: A Review, a Survey, and a Criticism,* Cambridge, MIT Press, 1978.

[R12.12] Colburn, T.R., *Philosophy and Computer Science,* Armonk, M.E. Sharpe, 2000.

[R12.13] Cover, T.M., Thomas, J.A.: *Elements of Information Theory,* New York, Wiley, 1991.

[R12.14] Dennett, D.C.: "Intentional Systems," *The Journal of Philosophy,* Vol. 68, 1971, pp. 87–106.

[R12.15] Deutsch, D., *The Fabric of Reality,* London Penguin, 1997.

[R12.16] Devlin, K.J., *Logic and Information.* Cambridge, Cambridge University Press, 1991.

[R12.17] Fetzer, J.H., "Information, Misinformation, and Disinformation" *Minds and Machines,* Vol. 14(2), 2004, pp. 223–229

[R12.18] Floridi, L., *Philosophy and Computing: An Introduction,* London, Routledge, (1999)

[R12.19] Floridi, L., "What Is the Philosophy of Information?" *Metaphilosophy,* Vol. 33(1–2) (2002), pp. 123–145.

[R12.20] Floridi, L., "Two Approaches to the Philosophy of Information," *Minds and Machines,* Vol. 13(4), (2003), pp. 459–469

[R12.21] Floridi, L.: "Open Problems in the Philosophy of Information," *Metaphilosophy* Vol. 35 (4), 2004, pp. 554–582.

[R12.22] Floridi, L., "Outline of a Theory of Strongly Semantic Information," *Minds and Machines,* Vol. 14(2), 2004, pp. 197–222.

[R12.23] Floridi, L., "Is Information Meaningful Data?," *Philosophy and Phenomenological Research,* Vol. 70(2), 2005, pp. 351–370.

[R12.24] Fox, C.J.: *Information and Misinformation: An Investigation of the Notions of Information, Misinformation, Informing, and Misinforming,* Westport Greenwood Press, 1983.

[R12.25] Frieden, B.R., *Science from Fisher Information: A Unification,* Cambridge University Press, Cambridge, 2004.

[R12.26] Golan, A., "Information and Entropy Econometrics - Editor's View". Journal of Econometrics, Vol. 107(1–2), 2002, pp. 1–15

[R12.27] Graham, G., *The Internet: A Philosophical Inquiry.,* London, Routledge, 1999.

[R12.28] Grice, H.P. *Studies in the Way of Words,* Cambridge, Harvard University Press, 1989.

[R12.29] Hanson, P.P. (ed.), *Information, language, and cognition*, Vancouver, University of British Columbia Press, 1990.

[R12.30] Harms, W.F., "The Use of Information Theory in Epistemology," *Philosophy of Science*, Vol. 65(3), 472–501 (1998).

[R12.31] Heil, J., "Levels of Reality," *Ratio* Vol. 16(3), 2003, pp. 205–221.

[R12.32] Hintikka, J., Suppes, P. (eds.), *Information and Inference*, Reidel, Dordrecht, 1970.

[R12.33] Kemeny, J., "A Logical Measure Function," *Journal of Symbolic Logic* Vol. 18, 1953, pp. 289–308.

[R12.34] Kolin K.K. "The Nature of Information and Philosophical Foundations of Informatics". *Open Education*, Vol. 2, (2005) pp. 43–51.

[R12.35] Kolin K.K. "The Evolution of Informatics," *Information Technologies*, Vol. 1, 2005, pp. 2–16.

[R12.36] Kolin K.K., "The Formation of Informatics as Basic Science and Complex Scientific Problems," In K. Kolin (Ed.), *Systems and Means of Informatics. Special Issue. Scientific and Methodological Problems of Informatics*. Moscow: IPI RAS, 2006, pp. 7–57.

[R12.37] Kolin K.K., Fundamental Studies in Informatics: A General Analysis, Trends and Prospects. *Scientific and Technical Information*, Vol. 1, (7), 2007, pp. 5–11.

[R12.38] Kolin K.K. "Structure of Reality and the Phenomenon of Information," *Open Education*, Vol. 5, 2008, pp. 56–61.

[R12.39] Losee, R.M., "A Discipline Independent Definition of Information," *Journal of the American Society for Information Science* Vol. 48(3), 1997, pp. 254–269.

[R12.40] Lozinskii, E. "Information and evidence in logic systems," *Journal of Experimental and Theoretical Artificial Intelligence*, Vol. 6, 1994, pp. 163–193.

[R12.41] Machlup, F., Mansfield, U. (eds.), *The Study of Information: Interdisciplinary Messages*, New York, Wiley, 1983.

[R12.42] MacKay, D.M., *Information, Mechanism and Meaning*, Cambridge MIT Press, 1969.

[R12.43] Marr, D., Vision: *A Computational Investigation into the Human Representation and Processing of Visual Information*. San Francisco, W.H. Freeman, 1982.

[R12.44] Mingers, J., "The Nature of Information and Its Relationship to Meaning," In: Winder, R.L., et al. (eds.) *Philosophical Aspects of Information Systems*, London, Taylor and Francis, 1997, pp. 73–84.

[R12.45] Nauta, D., *The Meaning of Information*, The Hague, Mouton, 1972.

[R12.46] Newell, A. "The Knowledge Level". *Artificial Intelligence* Vol. 18, 1982, pp. 87–127.

[R12.47] Newell, A., Simon, H.A., "Computer Science as Empirical Inquiry: Symbols and Search," *Communications of the ACM*, Vol. 19, 1976, pp. 113–126.

[R12.48] Pierce, J.R., *An Introduction to Information Theory: Symbols, Signals and Noise*, New York, Dover Publications, 1980.

[R12.49] Poli, R., "The Basic Problem of the Theory of Levels of Reality" *Axiomathes*, Vol. 12, 2001, pp. 261–283.

[R12.50] Sayre, K.M.: *Cybernetics and the Philosophy of Mind*. London, Routledge and Kegan Paul, (1976)

[R12.51] Simon, H.A., *The Sciences of the Artificial*, Cambridge, MIT Press, 1996.

[R12.52] Smokler, H., "Informational Content: A Problem of Definition," *The Journal of Philosophy*, Vol. 63(8), 1966, pp. 201–211.

[R12.53] Ursul A.D., *The Nature of the Information. Philosophical Essay*. Moscow: Politizdat, (1968).

[R12.54] Ursul A.D., *Information. Methodological Aspects*. Moscow: Nauka, 1971.

[R12.55] Ursul A.D., *Reflection and Information*. Moscow: Nauka (1973).

[R12.56] Ursul A.D. *The Problem of Information in Modern Science: Philosophical Essays*. Moscow: Nauka. (1975).

[R12.57] Ursul, A.D. "The Problem of the Objectivity of Information" in Libor Kubát and Jiři Zeman (eds.), *Entropy and Information in Science and Philosophy*, New York, Elsevier, 1975.

[R12.58] Weaver, W., "The Mathematics of Communication," *Scientific American* Vol. 181(1), 1949, pp. 11–15.

[R12.59] Winder, R.L., Probert, S.K., Beeson, I.A.: *Philosophical Aspects of Information Systems,* London, Taylor & Francis, 1997.

R13. Planning, Prescriptive Science and Information in Cost-Benefit Analysis Analytics

[R13.1] Alexander Ernest R., *Approaches to Planning,* Philadelphia, Pa. Gordon and Breach, 1992.

[R13.2] Bailey, J., *Social Theory for Planning,* London, Routledge and Kegan Paul, 1975.

[R13.3] Burchell R.W. and G. Sternlieb (eds.), *Planning Theory in the 1980's: A Search for Future Directions,* New Brunswick, N. J., Rutgers University Center for Urban and Policy Research, 1978.

[R13.4] Camhis, Marios, *Planning Theory and Philosophy,* London, Tavistock Publicationa, 1979.

[R13.5] Chadwick, G., *A Systems View of Planning,* Oxford, Pergamon, 1971.

[R13.6] Cooke, P., *Theories of Planning and Special Development,* London, Hutchinson, 1983.

[R13.7] Dompere, Kofi K., and Taresa Lawrence, "Planning," in Syed B Hussain, *Encyclopedia of Capitalism,* Vol. II, New York, Facts On File, Inc., 2004, pp. 649–653.

[R13.8] Dompere, Kofi K., *Social Goal-Objective Formation, Democracy and National Interest: A Theory of Political Economy under Fuzzy Rationality,* (Studies in Systems, Decision and Control, Vol. 4), New York, Springer, 2014.

[R13.9] Dompere, Kofi K., *Fuzziness, Democracy Control and Collective Decision-Choice System: A Theory on Political Economy of Rent-Seeking and Profit-Harvesting,* (Studies In Systems, Decision and Control, Vol. 5), New York, Springer, 2014.

[R13.10] Dompere, Kofi K., *The Theory of Aggregate Investment in Closed Economic Systems,* Westport, CT, Greenwood Press, 1999.

[R13.11] Dompere, Kofi K., *The Theory of Aggregate Investment and Output Dynamics in Open Economic Systems,* Westport, CT, Greenwood Press, 1999.

[R13.12] Faludi, A., *Planning Theory,* Oxford, Pergamon, 1973.

[R13.13] Faludi, A.(ed.), *A Reader in Planning Theory,* Oxford, Pergamon, 1973.

[R13.14] Harwood, E.C. (ed.), Reconstruction of Economics, American Institute For Economic Research, Great Barrington, Mass, 1955., Also in John Dewey and Arthur Bently, 'Knowing and the known', Boston, Beacon Press, 1949, p. 269.

[R13.15] Kickert, W.J.M., *Organization of Decision-Making A Systems-Theoretic Approach,* New York, North-Holland, 1980.

[R13.16] Knight, Frank H. *Risk, Uncertainty and Profit,* Chicago, University of Chicago Press, 1971.

[R13.17] Knight, Frank H. *On History and Method of Economics,* Chicago, University of Chicago Press, 1971.

[R13.18] Morgenstern, Oscar, *On the Accuracy of Economic Observations,* Princeton, Princeton Univ. Press. 1973.

R14. Possible-Actual Worlds and Information Analytics

[R14.1] Adams, Robert M., "Theories of Actuality," *Noûs,* Vol. 8, 1974, pp. 211–231.

[R14.2] Allen, Sture (ed.) *Possible Worlds in Humanities, Arts and Sciences,* Proceedings of Nobel Symposium, Vol. 65, New York, Walter de Gruyter Pub., 1989.

[R14.3] Armstrong, D. M., *A Combinatorial Theory of Possibility*. Cambridge University Press, 1989.

[R14.4] Armstrong, D.M *A World of States of Affairs*, Cambridge, Cambridge University Press, 1997.

[R14.5] Bell, J.S., "Six Possible Worlds of Quantum Mechanics" in Allen, Sture (Ed.) *Possible Worlds in Humanities, Arts and Sciences*, Proceedings of Nobel Symposium, Vol. 65, New York, Walter de Gruyter Pub., 1989. pp. 359–373.

[R14.6] Bigelow, John. "Possible Worlds Foundations for Probability", *Journal of Philosophical Logic*, 5 (1976), pp. 299–320.

[R14.7] Bradley, Reymond and Norman Swartz, *Possible World: An Introduction to Logic and its Philosophy*, Oxford, Bail Blackwell, 1997.

[R14.8] Castañeda, H.-N. "Thinking and the Structure of the World", *Philosophia*, 4 (1974), pp. 3–40.

[R14.9] Chihara, Charles S. *The Worlds of Possibility: Modal Realism and the Semantics of Modal Logic*, Clarendon, 1998.

[R14.10] Chisholm, Roderick. "Identity through Possible Worlds: Some Questions", *Noûs*, 1 1967, pp. 1–8; reprinted in Loux, *The Possible and the Actual.*

[R14.11] Divers, John, *Possible Worlds*, London: Routledge, 2002.

[R14.12] Forrest, Peter. "Occam's Razor and Possible Worlds", *Monist*, 65 (1982), pp. 456–64.

[R14.13] Forrest, Peter, and Armstrong, D. M. "An Argument Against David Lewis' Theory of Possible Worlds", *Australasian Journal of Philosophy*, 62 (1984), pp. 164–168.

[R14.14] Grim, Patrick, "There is No Set of All Truths", *Analysis*, Vol. 46, 1986, pp. 186–191.

[R14.15] Heller, Mark. "Five Layers of Interpretation for Possible Worlds", *Philosophical Studies*, 90 (1998), pp. 205–214.

[R14.16] Herrick, Paul, *The Many Worlds of Logic,*. Oxford: Oxford University Press, 1999.

[R14.17] Krips, H. "Irreducible Probabilities and Indeterminism", *Journal of Philosophical Logic*, Vol. 18, 1989, pp. 155–172.

[R14.18] Kuhn, Thomas S., "Possible Worlds in History of Science" in Allen, Sture (ed.) *Possible Worlds in Humanities, Arts and Sciences*, Proceedings of Nobel Symposium, Vol. 65, New York, Walter de Gruyter Pub., 1989. pp. 9–41.

[R14.19] Kuratowski, K. and Mostowski, A. *Set Theory: With an Introduction to Descriptive Set Theory*, New York: North-Holland, 1976.

[R14.20] Lewis, David, *On the Plurality of Worlds*, Oxford, Basil Blackwell, 1986.

[R10.21] Loux, Michael J. (ed.) *The Possible and the Actual: Readings in the Metaphysics of Modality*, Ithaca & London: Cornell University Press, 1979.

[R14.22] Parsons, Terence, *Nonexistent Objects*, New Haven, Yale University Press, 1980.

[R14.23] Perry, John, "From Worlds to Situations", *Journal of Philosophical Logic*, Vol. 15, 1986, pp. 83–107.

[R14.24] Rescher, Nicholas and Brandom, Robert. *The Logic of Inconsistency: A Study in Non-Standard Possible-World Semantics And Ontology*, Rowman and Littlefield, 1979.

[R14.25] Skyrms, Brian. "Possible Worlds, Physics and Metaphysics", *Philosophical Studies*, Vol. 30, 1976, pp. 323–32.

[R14.26] Stalmaker, Robert C. "Possible World", *Noûs*, Vol. 10, 1976, pp. 65–75.

[R14.27] Quine, W.V.O. *Word and Object*, M.I.T. Press, 1960.

[R14.28] Quine, W.V.O "Ontological Relativity", *Journal of Philosophy*, 65 (1968), pp. 185–212.

R15. Rationality, Information, Games, Conflicts and Exact Reasoning

[**R15.1**] Border, Kim, *Fixed Point Theorems with Applications to Economics and Game Theory*, Cambridge, Cambridge University Press, 1985.

[**R15.2**] Brandenburger, Adam, "Knowledge and Equilibrium Games," *Journal of Economic Perspectives*, Vol. 6, 1992, pp. 83–102.

[**R15.3**] Campbell, Richmond and Lanning Sowden, *Paradoxes of Rationality and Cooperation: Prisoner's Dilemma and Newcomb's Problem*, Vancouver, University of British Columbia Press, 1985.

[**R15.4**] Gates Scott and Brian Humes, *Games, Information, and Politics: Applying Game Theoretic Models to Political Science*, Ann Arbor, University of Michigan Press, 1996.

[**R15.5**] Gjesdal, Froystein, "Information and Incentives: The Agency Information Problem," *Review of Economic Studies*, Vol. 49, 1982, pp. 373–390.

[**R15.6**] Harsanyi, John, "Games with Incomplete Information Played by 'Bayesian' Players I: The Basic Model," *Management Science,* Vol. 14, 1967, pp. 159–182.

[**R15.7**] Harsanyi, John, "Games with Incomplete Information Played by 'Bayesian' Players II: Bayesian Equilibrium Points," *Management Science*, Vol. 14, 1968, pp. 320–334.

[**R15.8**] Harsanyi, John, "Games with Incomplete Information Played by 'Bayesian' Players III: The Basic Probability Distribution of the Game," *Management Science*, Vol. 14, 1968, pp. 486–502.

[**R15.9**] Harsanyi, John, *Rational Behavior and Bargaining Equilibrium in Games and Social Situations*, New York Cambridge University Press, 1977.

[**R15.10**] Krasovskii, N.N. and A.I. Subbotin, *Game-theoretical Control Problems*, New York, Springer, 1988.

[**R15.11**] Lagunov, V. N., *Introduction to Differential Games and Control Theory*, Berlin, Heldermann Verlag, 1985.

[**R15.12**] Maynard Smith, John, *Evolution and the Theory of Games,* Cambridge, Cambridge University Press, 1982.

[**R15.13**] Myerson, Roger, *Game Theory: Analysis of Conflict*, Cambridge, Mass. Harvard University Press, 1991.

[**R15.14**] Rapoport, Anatol and Albert Chammah, *Prisoner's Dilemma: A Study in Conflict and Cooperation*, Ann Arbor, University of Michigan Press, 1965.

[**R15.15**] Roth, Alvin E., "The Economist as Engineer: Game Theory, Experimentation, and Computation as Tools for Design Economics," *Econometrica*, Vol. 70, 2002, pp. 1341–1378.

[**R15.16**] Shubik, Martin, *Game Theory in the Social Sciences: Concepts and Solutions*, Cambridge, Mass., MIT Press, 1982.

R16. Social Sciences, Mathematics and the Problems of Exact and Inexact Information

[**R16.1**] Ackoff, R.L., *Scientific Methods: Optimizing Applied Research Decisions*, New York, John Wiley, 1962.

[**R16.2**] Angyal, A. "The Structure of Wholes," *Philosophy of Sciences*, Vol. 6, #1, 1939, pp. 23–37.

[**R16.3**] Bahm, A.J., "Organicism: The Philosophy of Interdependence" *International Philosophical Quarterly*, Vol. VII # 2, 1967.

[**R16.4**] Bealer, George, *Quality and Concept*, Oxford, Clarendon Press, 1982.

[**R16.5**] Black, Max, *Critical Thinking*, Englewood Cliffs, N.J., Prentice-Hall, 1952.

[R16.6] Brewer, Marilynn B., and Barry E Collins (eds.) *Scientific Inquiry and Social Sciences,* San Francisco, Ca, Jossey-Bass Pub., 1981.

[R16.7] Campbell, D.T., "On the Conflicts Between Biological and Social Evolution and Between Psychology and Moral Tradition", *American Psychologist,* Vol. 30, 1975, pp. 1103–1126.

[R16.8] Churchman, C. W. and P. Ratoosh (eds.) *Measurement: Definitions and Theories,* New York, John Wiley, 1959.

[R16.9] Foley, Duncan, "Problems versus Conflicts Economic Theory and Ideology" American Economic Association Papers and Proceedings, Vol. 65, May 1975, pp. 231–237.

[R16.10] Garfinkel, Alan, *Forms of Explanation: Structures of Inquiry in Social Science,* New Haven, Conn., Yale University Press, 1981.

[R16.11] Georgescu-Roegen, Nicholas, *Analytical Economics,* Cambridge, Harvard University Press, 1967.

[R16.12] Gillespie, C., *The Edge of Objectivity,* Princeton, Princeton University Press, 1960.

[R16.13] Hayek, F.A., *The Counter-Revolution of Science,* New York, Free Press of Glencoe Inc, 1952.

[R16.14] Laudan, L., *Progress and Its Problems: Towards a Theory of Scientific Growth,* Berkeley, CA, University of California Press, 1961.

[R16.15] Marx, Karl, *The Poverty of Philosophy,* New York, International Pub. 1971.

[R16.16] Phillips, Denis C., *Holistic Thought in Social Sciences,* Stanford, CA, Stanford University Press, 1976.

[R16.17] Popper, K., *Objective Knowledge,* Oxford, Oxford University Press, 1972.

[R16.18] Rashevsky, N. "Organismic Sets: Outline of a General Theory of Biological and Social Organism," *General Systems,* Vol XII, 1967, pp. 21–28.

[R16.19] Roberts, Blaine, and Bob Holdren, *Theory of Social Process,* Ames, Iowa University Press, 1972.

[R16.20] Rudner, Richard S., *Philosophy of Social Sciences,* Englewood Cliff, N.J., Prentice Hall, 1966.

[R16.21] Simon, H. A., "The Structure of Ill-Structured Problems," *Artificial Intelligence,* Vol. 4, 1973, pp. 181–201.

[R16.22] Toulmin, S., *Foresight and understanding: An Enquiry into the Aims of Science,* New York, Harper and Row, 1961.

[R16.23] Winch, Peter, *The Idea of a Social Science,* New York, Humanities Press, 1958.

R17. Tranformations, Decisions, Polarity, Duality and Conflict

[R17.1] Anovsky, 0mely M.E., *Linin and Modern Natural Science, Moscow,* Progress Pub. 1978.

[R17.2] Arrow, Kenneth J., "Limited Knowledge and Economic Analysis", American Economic Review, Vol. 64, 1974, pp. 1–10.

[R17.3] Berkeley, George, *Treatise Concerning the Principles of Human Knowledge, Works,* Vol. I (edited by A. Fraser), Oxford, Oxford University Press, 1871–1814.

[R17.4] Berkeley, George, "Material Things are Experiences of Men or God" in [R1.5], 1967, pp. 658–668.

[R17.5] Brody, Baruch A. (ed.), *Readings in the Philosophy of Science,* Englewood Cliffs, NJ., Prentice-Hall Inc., 1970.

[R17.6] Brouwer, L.E.J., "Consciousness, Philosophy, and Mathematics," in Benecerraf, P. and H. Putnam (eds.), *Philosophy of Mathematics: Selected Readings,* Cambridge, Cambridge University Press, 1983. pp. 90–96.

[R17.7] Brown, B. and J Woods (eds.), *Logical Consequence; Rival Approaches and New Studies in exact Philosophy: Logic, Mathematics and Science,* Vol. II Oxford, Hermes, 2000.

[R17.8] Cornforth, Maurice, *Dialectical Materialism and Science,* New York, International Pub. 1960.

[R17.9] Cornforth, Maurice, *Materialism and Dialectical Method*, New York, International Pub. 1953.

[R17.10] Cornforth, Maurice, *Science and Idealism: an Examination of "Pure Empiricism"*, New York International Pub. 1947.

[R17.11] Cornforth, Maurice, *The Open Philosophy and the Open Society: A Reply to Dr. Karl Popper's Refutations of Marxism* New York, International Pub. 1968.

[R17.12] Cornforth, Maurice, *The Theory of Knowledge*, New York, International Pub. 1960.

[R17.13] Dompere, Kofi K., "On Epistemology and Decision-Choice Rationality" in R. Trapple (ed.), *Cybernetics and System Research*, New York, North Holland, 1982, pp. 219–228.

[R17.14] Dompere, Kofi K. and M. Ejaz, *Epistemics of Development Economics: Toward a Methodological Critique and Unity*, Westport, CT, Greenwood Press, 1995.

[R17.15] Dompere, Kofi K., *The Theory of Categorial Conversion: Rational Foundations of Nkrumaism in socio-natural Systemicity and Complexity*, London, Adonis-Abbey Pubs. 2016–2017.

[R17.16] Dompere, Kofi K., *The Theory of Philosophical Consciencism: Practice Foundations of Nkrumaism in Social Systemicity*, London, Adonis-Abbey Pubs., 2016–2017.

[R17.17] Dompere, Kofi K., A General Theory of Information: Definitional Foundations and Critique of the Tradition, Working Monographs on Mathematics, Philosophy, Economic and Decision Theories, Washington, D.C., Department of Economics Howard University, 2016.

[R17.18] Dompere, Kofi K., *The Theory of Info-dynamics: Epistemic and Analytical Foundations*, Working Monographs on Mathematics, Philosophy, Economic and Decision Theories, Washington, D.C., Department of Economics, Howard University, 2016.

[R17.19] Dompere, Kofi K., *Polyrhythmicity: Foundations of African Philosophy*, London, Adonis and Abbey Pub, 2006.

[R17.20] Engels, Frederick, *Dialectics of Nature*, New York, International Pub., 1971.

[R17.21] Engels, Frederick, *Origin of the Family, Private Property and State*, New York, International Pub., 1971.

[R17.22] Ewing, A.C., "A Reaffirmation of Dualism" in [R1.5], pp. 454–461.

[R17.23] Fedoseyer, P.N. et al., *Philosophy in USSR: Problems of Dialectical Materialism*, Moscow, Progress Pub., 1977.

[R17.24] Kedrov, B. M., "On the Dialectics of Scientific Discovery," *Soviet Studies in Philosophy*, Vol. 6 1967, pp. 16–27.

[R17.25] Lenin, V. I. *Materialism and Empirio-Criticism: Critical Comments on Reactionary Philosophy*, New York, International Pub., 1970.

[R17.26] Lenin, V. I. *Collected Works Vol. 38: Philosophical Notebooks*, New York, International Pub., 1978.

[R17.27] Lenin, V. I., *On the National Liberation Movement*, Peking, Foreign Language Press, 1960.

[R17.28] Hegel, George, *Collected Works*, Berlin, Duncher und Humblot, 1832–1845 [also *Science of Logic*, translated by W. H. Johnston and L. G. Struther, London, 1951].

[R17.29] Hempel, Carl G. and P. Oppenheim, "Studies in the Logic of Explanation," in [R15.5], pp. 8–27.

[R17.30] Ilyenkov, E.V., *Dialectical Logic: Essays on its History and Theory*, Moscow, Progress Pub. 1977.

[R17.31] Keirstead, B.S., "The Conditions of Survival," American Economic Review, Vol. 40, #2, pp. 435–445.

[R17.32] Kühne, Karl, *Economics and Marxism, Vol. I: The Renaissance of the Marxian System*, New York, St Martin's Press, 1979.

[R17.33] Kühne, Karl, *Economics and Marxism, Vol. II: The Dynamics of the Marxian System*, New York, St Martin's Press, 1979.

[R17.34] March, J. C., "Bounded Rationality, Ambiguity and Engineering of Choice," *The Bell Journal of Economics*, Vol. 9 (2), 1978.

[R17.35] Marx, Karl, *Contribution to the Critique of Political Economy*, Chicago, Charles H. Kerr and Co. 1904.

[R17.36] Marx, Karl, *Economic and Philosophic Manuscripts of 1884*, Moscow, Progress Pub., 1967.

[R17.37] Marx, Karl, *The Poverty of Philosophy*, New York, International Publishers, 1963.

[R17.38] Marx, Karl, Economic and Philosophic Manuscripts of 1844, Moscow, Progress Pub, 1967.

[R17.39] Niebyl, Karl, H., "Modern Mathematics and Some Problems of Quantity, Quality and Motion in Economic Analysis," *Philosophy of Science*, Vol 7, # 1, January, 1940, pp. 103–120.

[R17.40] Price, H. H., *Thinking and Experience*, London, Hutchinson, 1953.

[R17.41] Putman, H., *Reason, Truth and History*, Cambridge, Cambridge University Press, 1981.

[R17.42] Putman. H., *Realism and Reason*, Cambridge, Cambridge University Press, 1983.

[R17.43] Robinson, Joan, *Economic Philosophy*. New York, Anchor Books, 1962.

[R17.44] Robinson, Joan, *Freedom and Necessity: An Introduction to the Study of Society*, New York, Vintage Books, 1971.

[R17.45] Robinson, Joan, *Economic Heresies: Some Old-Fashioned Questions in Economic Theory*, New York, Basic Books, 1973.

[R17.46] Schumpeter, Joseph A., *The Theory of Economic Development*, Cambridge, Mass. Harvard University Press, 1934.

[R17.47] Schumpeter, Joseph A., *Capitalism, Socialism and Democracy*, New York, Harper & Row, 1950.

[R17.48] Schumpeter, Joseph A., "March to Socialism," *American Economic Review*, Vol. 40 May 1950, pp. 446 456.

[R17.49] Schumpeter, Joseph A., "Theoretical Problems of Economic Growth" *Journal of Economic History* Vol. 8, Supplement, 1947, pp. 1–9.

[R17.50] Schumpeter, Joseph A., "The Analysis of Economic Change," *Review of Economic Statistics*, Vol. 17, 1935, pp. 2–10.

R18. Vagueness, Approximation and Reasoning in the Information-Knowledge Process

[R18.1] Adams, E. w., and H. F. Levine, "On the Uncertainties Transmitted from Premises to Conclusions in deductive Inferences," *Synthese* Vol. 30, 1975, pp. 429–460.

[R18.2] Arbib, M. A., *The Metaphorical Brain*, New York, McGraw-Hill, 1971.

[R18.3] Bečvář Jiří, "Notes on Vagueness and Mathematics," in Skala, Heinz J. et al., (eds.), *Aspects of Vagueness*, Dordrecht, D. Reidel Co. 1984, pp. 1–11.

[R18.4] Black, M, "Vagueness: An Exercise in Logical Analysis," *Philosophy of Science*, Vol. 17, 1970, pp. 141–164.

[R18.5] Black, M. "Reasoning with Loose Concepts," *Dialogue*, Vol. 2, 1973, pp. 1–12.

[R18.6] Black, Max, *Language and Philosophy*, Ithaca, N.Y.: Cornell University Press. 1949.

[R18.7] Black, Max, *The Analysis of Rules*, in Black, Max [R18.8] *Models and Metaphors: Studies in Language and Philosophy*, Ithaca, New York: Cornell University Press, 1962, pp. 95–139.

[R18.8] Black, Max, *Models and Metaphors: Studies in Language and Philosophy*, Ithaca, New York: Cornell University Press, 1962.

[R18.9] Black, Max *Margins of Precision*, Ithaca: Cornell University Press, 1970.

[R18.10] Boolos, G. S. and R. C. Jeffrey, *Computability and Logic*, New York, Cambridge University Press, 1989.

[R18.11] Cohen, P. R., *Heuristic Reasoning about uncertainty: An Artificial Intelligent Approach*, Boston, Pitman, 1985.

[R18.12] Darmstadter, H., "Better Theories," *Philosophy of Science*, Vol. 42, 1972, pp. 20–27.

[R18.13] Davis, M., *Computability and Unsolvability*, New York, McGraw-Hill, 1958.

[R18.14] Dummett, M., "Wang's Paradox," *Synthese,* Vol. 30, 1975, pp. 301–324.

[R18.15] Dummett, M., *Truth and Other Enigmas*, Cambridge, Mass. Harvard University Press, 1978.

[R18.16] Endicott, Timothy, *Vagueness in the Law*, Oxford, Oxford University Press, 2000.

[R18.17] Evans, Gareth, "Can there be Vague Objects?," *Analysis,* Vol. 38, 1978, p. 208.

[R18.18] Fine, Kit, "Vagueness, Truth and Logic," *Synthese,* Vol. 54, 1975, pp. 235–259.

[R18.19] Gale, S., "Inexactness, Fuzzy Sets and the Foundation of Behavioral Geography," *Geographical Analysis*, Vol. 4, #4, 1972, pp. 337–349.

[R18.20] Ginsberg, M. L. (ed.), *Readings in Non-monotonic Reason*, Los Altos, Ca., Morgan Kaufman, 1987.

[R18.21] Goguen, J. A., "The Logic of Inexact Concepts," *Synthese*, Vol. 19, 1968/69, pp. 325–373.

[R18.22] Grafe, W., "Differences in Individuation and Vagueness," in A. Hartkamper and H.-J. Schmidt, *Structure and Approximation in Physical Theories*, New York, Plenum Press, 1981. pp. 113–122.

[R18.23] Goguen, J. A, "The Logic of Inexact Concepts" *Synthese*, Vol. 19, 1968–1969.

[R18.24] Graff, Delia and Timothy (eds.), *Vagueness*, Aldershot, Ashgate Publishing, 2002.

[R18.25] A. Hartkämper and H.J. Schmidt (eds.), *Structure and Approximation in Physical Theories*, New York, Plenum Press, 1981.

[R18.26] Hersh, H.M. et. al., "A Fuzzy Set Approach to Modifiers and Vagueness in Natural Language," *J. Experimental*, Vol. 105, 1976, pp. 254–276.

[R18.27] Hilpinen, R., "Approximate Truth and Truthlikeness," in M. Pprelecki et al. (eds.) *Formal Methods in the Methodology of Empirical Sciences*, Wroclaw, Reidel, Dordrecht and Ossolineum, 1976, pp. 19–42.

[R18.28] Hockney D. et al. (eds.), *Contemporary Research in Philosophical Logic and Linguistic Semantics*, Dordrecht-Holland, Reidel Pub. Co. 1975.

[R18.29] Höhle Ulrich et al (eds.), *Non-Clasical Logics and their Applications to Fuzzy Subsets: A Handbook of the Mathematical Foundations of Fuzzy Set Theory*, Boston, Mass. Kluwer, 1995.

[R18.30] Katz, M., "Inexact Geometry," *Notre-Dame Journal of Formal Logic*, Vol. 21, 1980, pp. 521–535.

[R18.31] Katz, M., "Measures of Proximity and Dominance," *Proceedings of the Second World Conference on Mathematics at the Service of Man*, Universidad Politecnica de Las Palmas, 1982, pp. 370–377.

[R18.32] Katz, M., "The Logic of Approximation in Quantum Theory," *Journal of Philosophical Logic*, Vol. 11, 1982, pp. 215–228.

[R18.33] Keefe, Rosanna, *Theories of Vagueness*, Cambridge, Cambridge University Press, 2000.

[R18.34] Keefe, Rosanna and Peter Smith (eds.) *Vagueness: A Reader*, Cambridge, MIT Press, 1996.

[R18.35] Kling, R., "Fuzzy Planner: Reasoning with Inexact Concepts in a Procedural Problem-solving Language," *Jour. Cybernetics*, Vol. 3, 1973, pp. 1–16.

[R18.36] Kruse, R.E. et. al., *Uncertainty and Vagueness in Knowledge Based Systems: Numerical Methods*, New York, Springer, 1991.

[R18.37] Ludwig, G., "Imprecision in Physics," in A. Hartkämper and H.J. Schmidt (eds.), *Structure and Approximation in Physical Theories*, New York, Plenum Press, 1981, pp. 7–19.

[R18.38] Kullback, S. and R. A. Leibler, "Information and Sufficiency," *Annals of Math. Statistics*, Vol. 22, 1951, pp. 79–86.

[R18.39] Lakoff, George, "Hedges: A Study in Meaning Criteria and Logic of Fuzzy Concepts," in, Hockney D. et al. (eds.), *Contemporary Research in Philosophical Logic and Linguistic Semantics*, Dordrecht-Holland, Reidel Pub. Co. 1975, pp. 221–271.

[R18.40] Lakoff, G., "Hedges: A Study in Meaning Criteria and the Logic of Fuzzy Concepts," *Jour. Philos. Logic*, Vol. 2, 1973, pp. 458–508.

[R18.41] Levi, I., *The Enterprise of Knowledge*, Cambridge, Mass. MIT Press, 1980.

[R18.42] Łucasiewicz, J., *Selected Works: Studies in the Logical Foundations of Mathematics,* Amsterdam, North-Holland, 1970.

[R18.43] Machina, K.F., "Truth, Belief and Vagueness," *Jour. Philos. Logic,* Vol. 5, 1976, pp. 47–77.

[R18.44] Menges, G., et. al., "On the Problem of Vagueness in the Social Sciences," in Menges, G. (ed.), *Information, Inference and Decision,* D. Reidel Pub., Dordrecht, Holland, 1974, pp. 51–61.

[R18.45] Merricks, Trenton, "Varieties of Vagueness," *Philosophy and Phenomenological Research,* Vol. 53, 2001, pp. 145–157.

[R18.46] Mycielski, J., "On the Axiom of Determinateness," *Fund. Mathematics,* Vol. 53, 1964, pp. 205–224.

[R18.47] Mycielski, J., "On the Axiom of Determinateness II," *Fund. Mathematics,* Vol. 59, 1966, pp. 203–212.

[R18.48] Naess, A., "Towards a Theory of Interpretation and Preciseness," in L. Linsky (ed.) *Semantics and the Philosophy of Language,* Urbana, Ill. Univ. of Illinois Press, 1951.

[R18.49] Narens, Louis, "The Theory of Belief," *Journal of Mathematical Psychology,* Vol. 49, 2003, pp. 1–31.

[R18.50] Narens, Louis, "A Theory of Belief for Scientific Refutations," *Synthese,* Vol. 145, 2005, pp. 397–423.

[R18.51] Netto, A. B., "Fuzzy Classes," *Notices, Amar, Math. Society,* Vol. 68T- H28, 1968, p. 945.

[R18.52] Neurath, Otto et al. (eds.), *International Encyclopedia of Unified Science,* Vol. 1–10, Chicago, University of Chicago Press, 1955.

[R18.53] Niebyl, Karl, H., "Modern Mathematics and Some Problems of Quantity, Quality and Motion in Economic Analysis," *Science,* Vol 7, #1, January, 1940, pp. 103–120.

[R18.54] Orlowska, E., "Representation of Vague Information," *Information Systems,* Vol. 13, #2, 1988, pp. 167–174.

[R18.55] Parrat, L. G., *Probability and Experimental Errors in Science,* New York, John Wiley and Sons, 1961.

[R18.56] Raffman. D., "Vagueness and Context-sensitivity," *Philosophical Studies,* Vol. 81, 1996, pp. 175–192.

[R18.57] Reiss, S., *The Universe of Meaning,* New York, The Philosophical Library, 1953.

[R18.58] Russell, B., "Vagueness," *Australian Journal of Philosophy,* Vol. 1, 1923, pp. 84–92.

[R18.59] Russell, B., *An Inquiry into Meaning and Truth,* New York, Norton, 1940.

[R18.60] Shapiro, Stewart, *Vagueness in Context,* Oxford, Oxford University Press, 2006.

[R18.61] Skala, H. J. "Modelling Vagueness," in M. M. Gupta and E. Sanchez, *Fuzzy Information and Decision Processes,* Amsterdam North-Holland, 1982, pp. 101–109.

[R18.62] Skala, Heinz J. et al., (eds.), *Aspects of Vagueness,* Dordrecht, D. Reidel Co. 1984.

[R18.63] Sorensen, Roy, *Vagueness and Contradiction,* Oxford, Oxford University Press, 2001.

[R18.64] Tamburrini, G. and S. Termini, "Some Foundational Problems in Formalization of Vagueness," in M. M. Gupta et al (eds.), *Fuzzy Information and Decision Processes,* Amsterdam, North Holland, 1982, pp. 161–166.

[R18.65] Termini, S. "Aspects of Vagueness and Some Epistemological Problems Related to their Formalization," in Skala, Heinz J. et al., (eds.), *Aspects of Vagueness,* Dordrecht, D. Reidel Co. 1984, pp. 205–230.

[R18.66] Tikhonov, Andrey N. and Vasily Y. Arsenin., *Solutions of Ill-Posed Problems,* New York, John Wiley and Sons, 1977.

[R18.67] Tversky, A. and D. Kahneman, "Judgments under Uncertainty: Heuristics and Biases," *Science,* Vil 185 September 1974, pp. 1124–1131.

[R18.68] Ursul, A. D., "The Problem of the Objectivity of Information," in Kubát, Libor and J. Zeman (eds.), *Entropy and Information,* Amsterdam, Elsevier, 1975, pp. 187–230.

[R18.69] Vardi, M. (ed.), *Proceedings of Second Conference on Theoretical Aspects of Reasoning about Knowledge,* Asiloman, Ca, Los Altos, Ca, Morgan Kaufman, 1988.

[R18.70] Verma, R.R., "Vagueness and the Principle of the Excluded Middle," *Mind*, Vol. 79, 1970, pp. 66–77.

[R18.71] Vetrov, A. A., "Mathematical Logic and Modern Formal Logic," *Soviet Studies in Philosophy*, Vol. 3, #1, 1964, pp. 24–33.

[R18.72] von Mises, Richard, *Probability, Statistics and Truth,* New York, Dover Pub. 1981.

[R18.73] Williamson, Timothy, *Vagueness,* London, Routledge, 1994.

[R18.74] Wiredu, J.E., "Truth as a Logical Constant with an Application to the Principle of the Excluded Middle," *Philos. Quart.*, Vol. 25, 1975, pp. 305–317.

[R18.75] Wright, C., "On Coherence of Vague Predicates," *Synthese*, Vol. 30, 1975. pp. 325–365.

[R18.76] Wright, Crispin, "The Epistemic Conception of Vagueness," *Southern Journal of Philosophy,* Vol. 33, Supplement, 1995, pp. 133–159.

[R18.77] Zadeh, L. A., "A Theory of Commonsense Knowledge," in Skala, Heinz J. et al., (eds.), *Aspects of Vagueness*, Dordrecht, D. Reidel Co. 1984., pp. 257–295.

[R18.78] Zadeh, L. A., "The Concept of Linguistic Variable and its Application to Approximate reasoning," *Information Science*, Vol. 8, 1975, pp. 199–249 (Also in Vol. 9, pp. 40–80).

R19. Vagueness, Disinformation, Misinformation and Fuzzy Game Theory in Socio-Natural Transformations

[R19.1] Aubin, J.P. "Cooperative Fuzzy Games", Mathematics of Operations Research, Vol. 6, 1981, pp. 1–13.

[R19.2] Aubin, J.P. Mathematical Methods of Game and Economics Theory, New York, North Holland. 1979.

[R19.3] Butnaria, D., "Fuzzy Games: A description pf the concepts," Fuzzy sets and systems Vol. 1, 1978, pp. 181–192.

[R19.4] Butnaria, D., "Stability and shapely value for a n-persons Fuzzy Games," Fuzzy sets and systems, Vol. 4, #1, 1980, pp. 63–72.

[R19.5] Nurmi, H.., "A Fuzzy Solution to a Majority Voting Game," Fuzzy sets and systems, Vol. 5, 1981, pp. 187–198.

[R19.6] Regade., R. K., "Fuzzy Games in the Analysis of Options," jour. Of Cybernetics, Vol. 6, 1976, pp. 213–221.

[R19.7] Spillman, B. et al., "Coalition Analysis with Fuzzy Sets," Kybernetes, Vol. 8, 1979, pp. 203–211.

[R19.8] Wernerfelt, B., "Semifuzzy Games" Fuzzy sets and systems, Vol. 19, 1986, pp. 21–28.

R20. Weapon Foundations for Information System

[R20.1] Forte, B., "On a System of Functional Equation in Information Theory," *Aequationes Math.* Vol 5, 1970, pp. 202–211.

[R20.2] Gallick, James, *The Information: A History, a Theory, a Flood.* Pantheon, New York, NY, 2011.

[R20.3] Hopcroft, John, E. Rajeev Motwani and Ljeffrey D. Ullman, *Introduction to automata Theory, Languages, and Computation*, Pearson Education, 2000.

[R20.4] Howard, N., *Paradoxes of Rationality*, Cambridge, Mass., MIT Press, 1972.

[R20.5] Ingarden, R.S, "A Simplified Axiomatic Definition of Information," *Bull. Acad. Polo Sci. Ser, Sci. Math Astronomy Phys.*, Vol. 11, 1963, pp. 209–212.

[R20.6] Lee, P. M., "On the Axioms of Information Theory," *Annals of Math. Statistics*, vol. 35, 1964, pp. 415–418.

[R20.7] Luce, R. D. (ed.), *Development in Mathematical Psychology*, Westport, Greenwood Press, 1960.

[R20.8] Luciano Floridi, "Is Information Meaningful Data?," *Philosophy and Phenomenological Research,* 70 (2), 2005, pp. 351–370.

[R20.9] Meyer, L, "Meaning in Music and Information Theory," *Journal of Aesthetics and Art Criticism,* Vol. 15, 1957, pp. 412–424.

[R20.10] Rich, Elaine, *Automata, Computability, and Complexity: Theory and Applications,* Peason, 2008.

[R20.11] Shannon, Claude E., "The Mathematical Theory of Communication," *Bell System Technical Journal,* Vol. 27,#3, 1945, pp. 379–423 and Vol. 27, #4, 1948, pp. 623–666.

[R20.12] *Shannon, Claude E.* and Warren Weaver, *The Mathematical Theory of Communication.* University of Illinois Press, 1949.

[R20.13] Theil, Henri, *Statistical Decomposition Analysis,* Amsterdam, North-Holland, 1974.

[R20.14] *Vigo, R.* "Representational information: a new general notion and measure of Information". *Information Sciences.* Vol. 181, (2011), pp. 4847–4859.

[R20.15] *Vigo, R* "Complexity over Uncertainty in Generalized Representational Information Theory (GRIT): A Structure-Sensitive General Theory of Information". *Information.* 4(1), 2013, pp. 1–30.

[R20.16] Vigo, R. *Mathematical Principles of Human Conceptual Behavior: The Structural Nature of Conceptual Representation and Processing,* Routledge, New York and London, 2014.

[R20.17] Wicker Stephen B. and Saejoon Kim, *Fundamentals of Codes, Graphs, and Iterative Decoding. Springer,* New York, 2003.

[R20.18] Wiener, N., Cybernetics, New York, John Wiley and Sons, 1948.

[R20.19] Wiener, N., The Human use of Human Beings, Boston, Mass, Houghton, 1950.

[R20.20] Young, Paul. *The Nature of Information,* Greenwood Publishing Group, Westport, Ct, 1987.

R21. Written and Audio Languages and Information

[R21.1] Agha, Agha *Language and Social Relations,* Cambridge, Cambridge University Press, 2006.

[R21.2] Aitchison, Jean (ed.), *Language Change: Progress or Decay?* Cambridge, New York, Melbourne: Cambridge University Press, 2001.

[R21.3] Allerton, D. J. "Language as Form and Pattern: Grammar and its Categories". In Collinge, N.E. (ed.) *An Encyclopedia of Language.* London and New York: Routledge, 1989.

[R21.4] Anderson, Stephen, *Languages: A Very Short Introduction.* Oxford: Oxford University Press, 2012.

[R21.5] Aronoff, Mark; Fudeman, Kirsten, *What is Morphology.* New York, John Wiley & Sons, 2011.

[R21.6] Barber Alex & Robert J Stainton (eds.), *Concise Encyclopedia of Philosophy of Language and Linguistics.* New York, Elsevier, 2010.

[R21.7] Bauer, Laurie (ed.), *Introducing linguistic morphology* Washington, D.C., Georgetown University Press, 2003.

[R21.8] Brown, Keith; Ogilvie, Sarah, (eds.), *Concise Encyclopedia of Languages of the World,* New York Elsevier Science, 2008.

[R21.9a] Campbell, Lyle (ed.), *Historical Linguistics: an Introduction,* Cambridge, MA: MIT Press, 2004.

[R21.9b] Chao, Yuen Ren, *Language and Symbolic Systems,* Cambridge, Cambridge University Press. 1968.

[R21.10] Chomsky, Noam, *Syntactic Structures. The Hague: Mouton,* 1957.

[R21.11] Chomsky, Noam, *The Architecture of Language.* Oxford: Oxford University Press, 2000.

[R21.12] Clarke, David S., *Sources of semiotic: readings with commentary from antiquity to the present.* Carbondale: Southern Illinois University Press, 1990.

[R21.13] Comrie, Bernard (ed.), *Language universals and linguistic typology: Syntax and morphology.* Oxford: Blackwell, 1989.

[R21.14] Comrie, Bernard, (ed.), *The World's Major Languages.* New York: Routledge, 2009.

[R21.15] Coulmas, Florian, *Writing Systems: An Introduction to Their Linguistic Analysis.* Cambridge University Press, 2002.

[R21.16] Croft, William, Cruse, D. Alan, *Cognitive Linguistics. Cambridge, Cambridge University Press,* 2004.

[R21.17] Crystal, David, *The Cambridge Encyclopedia of Language.* Cambridge: Cambridge University Press, 1997.

[R21.18] Deacon, Terrence, *The Symbolic Species: The Co-evolution of Language and the Brain,* New York: W.W. Norton & Company, 1997.

[R21.19] Devitt, Michael and Sterelny, Kim *Language and Reality: An Introduction to the Philosophy of Language. Boston:* MIT Press, 1999.

[R21.20] Evans, Nicholas; Levinson, Stephen C. "The myth of language universals: Language diversity and its importance for cognitive science". *Behavioral and Brain Sciences* **32** (5), 2009, pp. 429–492.

[R21.21] Fitch, W. Tecumseh, *The Evolution of Language. Cambridge: Cambridge University Press,* 2010.

[R21.22] Foley, William A., *Anthropological Linguistics: An Introduction,* Oxford, Blackwell, 1997.

[R21.23] Greenberg, Joseph, *Language Universals: With Special Reference to Feature Hierarchies. The Hague: Mouton & Co.* 1966.

[R21.24] Hauser, Marc D.; Chomsky, Noam; Fitch, W. Tecumseh, "The Faculty of Language: What Is It, Who Has It, and How Did It Evolve?" *Science 22* **298,** (5598), 2002, pp. 1569–1579.

[R21.25] International Phonetic Association, *Handbook of the International Phonetic Association: A guide to the use of the International Phonetic Alphabet.* Cambridge: Cambridge University Press, 1999.

[R21.26] Katzner, Kenneth, *The Languages of the World,* New York: Routledge, 1999.

[R21.27] Labov, William, *Principles of Linguistic Change vol. I: Internal Factors,* Oxford, Blackwell, 1994.

[R21.28] Labov, William, *Principles of Linguistic Change vol. II: Social Factors,* Oxford, Blackwell, 2001.

[R21.29] Ladefoged, Peter, Maddieson, Ian, *The sounds of the world's languages,* Oxford: Blackwell. pp. 329–330, 1996.

[R21.30] Levinson, Stephen C. *Pragmatics.* Cambridge: Cambridge University Press, 1983.

[R21.31] Lewis, M. Paul (ed.), *Ethnologue: Languages of the World,* Dallas, Tex.: SIL International, 2009.

[R21.32] Lyons, John, *Language and Linguistics,* Cambridge, Cambridge University Press, 1981.

[R21.33] MacMahon, *April M.S., Understanding Language Change,* Cambridge, Cambridge University Press, 1994.

[R21.34] Matras, Yaron; Bakker, Peter, (eds). *The Mixed Language Debate: Theoretical and Empirical Advances.* Berlin: Walter de Gruyter, 2003.

[R21.35] Moseley, Christopher (ed.), *Atlas of the World's Languages in Danger,* Paris: UNESCO Publishing, 2010.

[R21.36] Nerlich, B. "History of pragmatics". In L. Cummings (ed.), *The Pragmatics Encyclopedia.* New York: Routledge, 2010, pp. 192–93.

[R21.37] Newmeyer, Frederick J., *The History of Linguistics.,* Linguistic Society of America, 2005.

[R21.38] Newmeyer, Frederick J., *Language Form and Language Function* (PDF). Cambridge, MA: MIT Press, 1998.

[R21.39] Nichols, Johanna *Linguistic diversity in space and time.* Chicago: University of Chicago Press, 1992.

[R21.40] Nichols, Johanna "Functional Theories of Grammar". *Annual Review of Anthropology,* Vol. **13**, 1984, pp. 97–117.

[R21.38] Senft, Gunter, *Systems of Nominal Classification.* Cambridge University Press. (ed.), 2008.

[R21.39] Swadesh, Morris, "The phonemic principle", *Language,* Vol. **10** (2): (1934), 117–129.

[R21.40] Tomasello, Michael "The Cultural Roots of Language". In B. Velichkovsky and D. Rumbaugh (Eds.), *Communicating Meaning: The Evolution and Development of Language.* Psychology Press, 1996, pp. 275–308.

[R21.41] Tomasello, Michael, *Origin of Human Communication.* MIT Press, 2008.

[R21.42] Ulbaek, Ib, "The Origin of Language and Cognition", In J. R. Hurford & C. Knight (eds.). *Approaches to the evolution of language.* Cambridge University Press. 1998, pp. 30–43.

Printed in the United States
By Bookmasters